WORKSHEETS

DEANA RICHMOND

BEGINNING AND INTERMEDIATE ALGEBRA
THIRD EDITION

Gary Rockswold
Minnesota State University, Mankato

Terry Krieger
Rochester Community and Technical College

PEARSON

Boston   Columbus   Indianapolis   New York   San Francisco   Upper Saddle River
Amsterdam   Cape Town   Dubai   London   Madrid   Milan   Munich   Paris   Montreal   Toronto
Delhi   Mexico City   Sao Paulo   Sydney   Hong Kong   Seoul   Singapore   Taipei   Tokyo

ISBN-13: 978-0-321-75661-9
ISBN-10: 0-321-75661-4

www.pearsonhighered.com

Contents

Chapter 1   1

Chapter 2   37

Chapter 3   57

Chapter 4   87

Chapter 5   103

Chapter 6   127

Chapter 7   149

Chapter 8   181

Chapter 9   207

Chapter 10   219

Chapter 11   253

Chapter 12   277

Chapter 13   303

Chapter 14   315

Answer Section   329

Chapter 1 Introduction to Algebra
1.1 Numbers, Variables, and Expressions

Natural Numbers and Whole Numbers ~ Prime Numbers and Composite Numbers ~ Variables, Algebraic Expressions, and Equations ~ Translating Words to Expressions

---

STUDY PLAN

Read: Read Section 1.1 on pages 2-8 in your textbook or eText.

Practice: Do your assigned exercises in your ☐ Book ☐ MyMathLab ☐ Worksheets

Review: Keep your corrected assignments in an organized notebook and use them to review for the test.

Key Terms

Exercises 1-10: Use the vocabulary terms listed below to complete each statement.
Note that some terms or expressions may not be used.

product	algebraic expression
equation	whole number
formula	composite number
variable	natural number
factor	prime number
	prime factorization

1. A(n) _____ is a special type of equation that expresses a relationship between two or more quantities.

2. The set of _____(s) may be expressed as 0, 1, 2, 3, 4, 5, 6, ...

3. A natural number greater than 1 that has only itself and 1 as natural number factors is a(n) _____.

4. Multiplying natural numbers 3 and 4 results in 12, a natural number. The result 12 is called the _____ and the numbers 3 and 4 are _____(s) of 12.

5. The expression $2 \cdot 2 \cdot 2 \cdot 3 \cdot 5$ is the _____ of 120.

6. A(n) _____ is a symbol, typically an italic letter such as x, y, z or F, used to represent an unknown quantity.

7. The set of _____(s) comprise the counting numbers and may be expressed as 1, 2, 3, 4, 5, 6, . . .

8. A(n) _____ consists of numbers, variables, operation symbols such as $+$, $-$, $\cdot$, and $\div$, and grouping symbols such as parentheses.

9. A natural number greater than 1 that has factors other than itself and 1 is a(n) _____ .

10. A(n) _____ is a mathematical statement that two algebraic expressions are equal.

Prime Numbers and Composite Numbers

Exercises 1-4: Refer to Example 1 on page 4 in your text and the Section 1.1 lecture video.

Classify each number as prime or composite, if possible. If a number is composite, write it as a product of prime numbers.

1. 29   1. _____

2. 0   2. _____

3. 65   3. _____

4. 180   4. _____

Variables, Algebraic Expressions, and Equations

Exercises 5-14: Refer to Examples 2-4 on pages 5-6 in your text and the Section 1.1 lecture video.

Evaluate each algebraic expression for the given value of x.

   5. $x + 6$; $x = 4$ **5.** _____

   6. $4x$; $x = 3$ **6.** _____

   7. $17 - x$; $x = 2$ **7.** _____

   8. $\dfrac{x}{(x-4)}$; $x = 8$ **8.** _____

Evaluate each algebraic expression for the given values of y and z.

   9. $4yz$; $y = 1,\ z = 5$ **9.** _____

   10. $z - y$; $y = 7,\ z = 1$ **10.** _____

   11. $\dfrac{z}{y}$; $y = 6,\ z = 18$ **11.** _____

Find the value of y for the given value(s).

   12. $y = x + 2$; $x = 13$ **12.** _____

   13. $y = \dfrac{x}{5}$; $x = 5$ **13.** _____

   14. $y = 4xz$; $x = 2,\ z = 3$ **14.** _____

Translating Words to Expressions

Exercises 15-20: Refer to Examples 5-7 on pages 6-7 in your text and the Section 1.1 lecture video.

Translate each phrase to an algebraic expression. Specify what each variable represents.

15. Five less than a number

15. _____

16. Three times the cost of a DVD

16. _____

17. A number minus 5, all multiplied by a different number

17. _____

18. The quotient of 100 and a number

18. _____

19. For each year after 2000, the average price of a gallon of milk has increased by about $0.15 per year.

   (a) What was the total increase in the price of a gallon of milk after 10 years, or in 2010?

   19. (a)_____

   (b) Write a formula (or equation) that gives the total increase P in the average price of a gallon of milk t years after 2000.

   (b)_____

   (c) Use your formula to estimate the total increase in the price of a gallon of milk in 2020.

   (c)_____

20. The volume V of a rectangular box equals its length l times its width w times its height h.

   (a) Write a formula that shows the relationship between these four quantities.

   20. (a)_____

   (b) Find the volume of a rectangular box that is 12 inches long, 5 inches wide, and 3 inches high.

   (b)_____

Chapter 1 Introduction to Algebra
1.2 Fractions

Basic Concepts ~ Simplifying Fractions to Lowest Terms ~ Multiplication and Division of Fractions ~ Addition and Subtraction of Fractions ~ An Application

---

STUDY PLAN

Read: Read Section 1.2 on pages 11-23 in your textbook or eText.

Practice: Do your assigned exercises in your ☐ Book ☐ MyMathLab ☐ Worksheets

Review: Keep your corrected assignments in an organized notebook and use them to review for the test.

Key Terms
Exercises 1-5: Use the vocabulary terms listed below to complete each statement.
Note that some terms or expressions may not be used.

> **reciprocal**
> **greatest common factor (GCF)**
> **basic principle of fractions**
> **multiplicative inverse**
> **lowest terms**
> **least common denominator (LCD)**

1. A fraction is said to be in _____ when the numerator and denominator have no factors in common.

2. The _____ for two or more fractions is the smallest number that is divisible by every denominator.

3. The _____ states that the value of a fraction is unchanged if the numerator and denominator of the fraction are multiplied (or divided) by the same nonzero number.

4. The _____ or _____ of a nonzero number a is $\frac{1}{a}$.

5. The _____ of two or more numbers is the largest factor that is common to those numbers.

Basic Concepts

Exercises 1-3: Refer to Example 1 on page 12 in your text and the Section 1.2 lecture video.

Give the numerator and denominator of each fraction.

1. $\dfrac{7}{15}$

1. _____

_____

2. $\dfrac{x}{yz}$

2. _____

_____

3. $\dfrac{a+2}{b-3}$

3. _____

_____

Simplifying Fractions to Lowest Terms

Exercises 4-7: Refer to Examples 2-3 on pages 13-14 in your text and the Section 1.2 lecture video.

Find the greatest common factor (GCF) for each pair of numbers.

4.   28, 70

4. _____

5.   32, 40

5. _____

Simplify each fraction to lowest terms.

6. $\dfrac{18}{45}$

6. _____

7. $\dfrac{45}{105}$

7. _____

Multiplication and Division of Fractions

Exercises 8-20: Refer to Examples 4-8 on pages 15-18 in your text and the Section 1.2 lecture video.

Multiply. Simplify the result when appropriate.

8. $\dfrac{3}{5} \cdot \dfrac{4}{7}$

8. _____

9. $\dfrac{6}{7} \cdot \dfrac{2}{3}$

9. _____

10. $4 \cdot \dfrac{5}{8}$

10. _____

11. $\dfrac{a}{b} \cdot \dfrac{c}{5}$

11. _____

Find each fractional part.

12. Three-fourths of one-third

12. _____

13. Two-thirds of one-fifth

13. _____

14. Three-fourths of eight

14. _____

15. Approximately nine-tenths of Team in Training participants are female. About one-third of those are over the age of 40. What fraction of Team in Training participants are women over the age of 40?

15. _____

Divide. Simplify the result when appropriate.

16. $\dfrac{1}{4} \div \dfrac{4}{7}$ 16. _____

17. $\dfrac{2}{3} \div \dfrac{2}{3}$ 17. _____

18. $6 \div \dfrac{12}{5}$ 18. _____

19. $\dfrac{x}{3} \div \dfrac{z}{y}$ 19. _____

20. Describe a problem for which the solution could be found by   20. _____
     dividing 3 by $\dfrac{1}{4}$.

Addition and Subtraction of Fractions

Exercises 21-29: Refer to Examples 9-12 on pages 19-22 in your text and the Section 1.2 lecture video.

Add or subtract as indicated. Simplify your answer to lowest terms when appropriate.

21. $\dfrac{5}{11} + \dfrac{7}{11}$ 21. _____

22. $\dfrac{13}{9} - \dfrac{7}{9}$ 22. _____

Find the LCD for each set of fractions.

23. $\dfrac{2}{3}, \dfrac{1}{4}$ 23. _____

24. $\dfrac{4}{15}, \dfrac{3}{20}$ 24. _____

Rewrite each set of fractions using the LCD.

25. $\dfrac{2}{3}, \dfrac{1}{4}$ 25. _____

26. $\dfrac{4}{15}, \dfrac{3}{20}$ 26. _____

Add or subtract as indicated. Simplify your answer to lowest terms when appropriate.

27. $\dfrac{2}{3} + \dfrac{1}{4}$ 27. _____

28. $\dfrac{4}{15} + \dfrac{3}{20}$ 28. _____

29. $\dfrac{1}{3} + \dfrac{2}{7} + \dfrac{5}{14}$ 29. _____

An Application

30.   A board measures $27\dfrac{1}{2}$ inches and needs to be cut into four   30. _____

         equal parts. Find the length of each piece.

Chapter 1 Introduction to Algebra
1.3 Exponents and Order of Operations

Natural Number Exponents ~ Order of Operations ~ Translating Words to Expressions

---

STUDY PLAN

Read: Read Section 1.3 on pages 26-32 in your textbook or eText.

Practice: Do your assigned exercises in your ☐ Book ☐ MyMathLab ☐ Worksheets

Review: Keep your corrected assignments in an organized notebook and use them to review for the test.

Key Terms
Exercises 1-3: Use the vocabulary terms listed below to complete each statement. Note that some terms or expressions may not be used. Some terms may be used more than once.

base	addition
parentheses	division
subtraction	absolute value
exponent	exponential expression
multiplication	

1. The _____ b^n, where n is a natural number, means $b^n = b \cdot b \cdot b \cdot \ldots \cdot b$ (with n factors).

2. In the expression b^n, the _____ is b and the _____ is n.

3. ORDER OF OPERATIONS
    Use the following order of operations. First perform all calculations within _____ and _____(s), or above and below the fraction bar.
    Evaluate all _____(s).
    Do all _____ and _____ from left to right.
    Do all _____ and _____ from left to right.

Natural Number Exponents

Exercises 1-9: Refer to Examples 1-3 on pages 28-29 in your text and the Section 1.3 lecture video.

Write each product as an exponential expression.

1. $6 \cdot 6 \cdot 6 \cdot 6 \cdot 6$

    1. _____

2. $\dfrac{1}{3} \cdot \dfrac{1}{3} \cdot \dfrac{1}{3} \cdot \dfrac{1}{3}$

    2. _____

3. $a \cdot a \cdot a \cdot a \cdot a \cdot a \cdot a$

    3. _____

Evaluate each expression.

4. 4^3

    4. _____

5. 10^5

    5. _____

6. $\left(\dfrac{2}{3}\right)^3$

    6. _____

Use the given base to write each number as an exponential expression. Check your results with a calculator, if one is available.

7. 1000 (base 10)

    7. _____

8. 32 (base 2)

    8. _____

9. 81 (base 3)

    9. _____

Order of Operations

Exercises 10-17: Refer to Examples 5-6 on pages 30-31 in your text and the Section 1.3 lecture video.

Evaluate each expression by hand.

10. $12 - 7 - 2$ 10. _____

11. $15 - (8 - 1)$ 11. _____

12. $4 + \dfrac{9}{3}$ 12. _____

13. $\dfrac{2+1}{7+8}$ 13. _____

14. $30 - 6 \cdot 4$ 14. _____

15. $6 + 5 \cdot 2 - (7 - 2)$ 15. _____

16. $\dfrac{2 + 2^3}{25 - 15}$ 16. _____

17. $4 \cdot 3^2 - (6 + 2)$ 17. _____

Translating Words to Expressions

Exercises 18-21: Refer to Example 7 on page 31 in your text and the Section 1.3 lecture video.

Translate each phrase into a mathematical expression and then evaluate it.

18.   Three cubed minus eight   **18.** _____

19.   Thirty plus four times two   **19.** _____

20.   Four cubed divided by two squared   **20.** _____

21.   Forty divided by the quantity ten minus two   **21.** _____

Chapter 1 Introduction to Algebra
1.4 Real Numbers and the Number Line

Signed Numbers ~ Integers and Rational Numbers ~ Square Roots ~ Real and Irrational Numbers ~ The Number Line ~ Absolute Value ~ Inequality

---

STUDY PLAN

Read: Read Section 1.4 on pages 34-42 in your textbook or eText.

Practice: Do your assigned exercises in your ☐ Book ☐ MyMathLab ☐ Worksheets

Review: Keep your corrected assignments in an organized notebook and use them to review for the test.

Key Terms
Exercises 1-10: Use the vocabulary terms listed below to complete each statement.
Note that some terms or expressions may not be used.

origin	average
rational number	additive inverse
principal square root	irrational number
integer	square root
absolute value	real number
opposite	

1. A real number that cannot be expressed by a fraction is a(n) _____.

2. If a is a positive number, then the _____ of a, denoted $\sqrt{a}$, is the positive square root of a.

3. The point on the number line associated with the real number 0 is called the _____.

4. The _____ of a set of numbers is found by adding the numbers and then dividing by how many numbers there are in the set.

5. A(n) _____ is any number that can be expressed as a ratio of two integers, $\frac{p}{q}$, where $q \neq 0$.

6. The _____, or _____, of a number a is $-a$.

7. The number b is a _____ of a number a if $b \cdot b = a$.

8. The _____ of a real number equals its distance on the number line from the origin.

9. The _____(s) include the natural numbers, zero, and the opposites of the natural numbers.

10. If a number can be represented by a decimal number, then it is a(n) _____.

Signed Numbers

Exercises 1-4: Refer to Examples 1-2 on page 35 in your text and the Section 1.4 lecture video.

Find the opposite of each expression.

1. 11

1. _____

2. $-\dfrac{3}{8}$

2. _____

3. $-(-5)$

3. _____

4. Find the additive inverse of $-b$, if $b = -\dfrac{1}{5}$.

4. _____

Integers and Rational Numbers

Exercises 5-8: Refer to Example 3 on page 36 in your text and the Section 1.4 lecture video.

Classify each number as one or more of the following: natural number, whole number, integer, or rational number.

5. $\dfrac{15}{3}$ 5. _____

6. -4 6. _____

7. 0 7. _____

8. $-\dfrac{8}{3}$ 8. _____

Square Roots

Exercises 9-11: Refer to Example 4 on page 37 in your text and the Section 1.4 lecture video.

Evaluate each square root. Approximate your answer to three decimal places when appropriate.

9. $\sqrt{64}$ 9. _____

10. $\sqrt{400}$ 10. _____

11. $\sqrt{6}$ 11. _____

Real and Irrational Numbers

Exercises 12-16: Refer to Examples 5-6 on pages 38-39 in your text and the Section 1.4 lecture video.

Classify each number as one or more of the following: natural number, whole number, integer, rational number, or irrational number.

12. $-\sqrt{7}$ 12. _____

13. -2 13. _____

14. $\dfrac{9}{14}$ 14. _____

15. $\sqrt{25}$ 15. _____

16. The table lists the height (in inches) of four students. Find the   16. _____
average height. Is the result a natural number, a rational number,
or an irrational number?   _____

Student	1	2	3	4
Height (in inches)	71	68	72	65

The Number Line

Exercises 17-19: Refer to Example 7 on page 39 in your text and the Section 1.4 lecture video.

Plot each real number on a number line.

17. $\dfrac{3}{4}$

```
← | | | | | | | | | | | | | | | | | | | | | →
  -10 -8  -6  -4  -2  0   2   4   6   8   10
```

18. $\sqrt{5}$

```
← | | | | | | | | | | | | | | | | | | | | | →
  -10 -8  -6  -4  -2  0   2   4   6   8   10
```

19. $-\pi$

```
← | | | | | | | | | | | | | | | | | | | | | →
  -10 -8  -6  -4  -2  0   2   4   6   8   10
```

Absolute Value

Exercises 20-22: Refer to Example 8 on page 40 in your text and the Section 1.4 lecture video.

Evaluate each expression.

20. $|6.7|$ 20. _____

21. $|-4|$ 21. _____

22. $|3-10|$ 22. _____

Inequality

Exercise 23: Refer to Example 9 on page 41 in your text and the Section 1.4 lecture video.

23. List the following numbers from least to greatest.   23. _____
     Then plot these numbers on a number line.

     $\pi, -\sqrt{3}, 0, 2.2, -3$

```
← | | | | | | | | | | | | | →
  -6   -4   2   0   2   4   6
```

Chapter 1 Introduction to Algebra
1.5 Addition and Subtraction of Real Numbers

Addition of Real Numbers ~ Subtraction of Real Numbers ~ Applications

---

STUDY PLAN

Read: Read Section 1.5 on pages 45-49 in your textbook or eText.

Practice: Do your assigned exercises in your ☐ Book ☐ MyMathLab ☐ Worksheets

Review: Keep your corrected assignments in an organized notebook and use them to review for the test.

Key Terms
Exercises 1-7: Use the vocabulary terms listed below to complete each statement.
Note that some terms or expressions may not be used.

sum	greater than
addends	difference
less than	less than or equal to
greater than or equal to	approximately equal
$a > b$	$a \geq b$
$a \leq b$	$a < b$

1. We say that a is _____ b, denoted _____, if either $a < b$ or $a = b$ is true.

2. The answer to a subtraction problem is the _____.

3. If a real number a is located to the right of a real number b on the number line, we say that a is _____ b and we write _____.

4. The symbol $\approx$ is used to represent _____.

5. We say that a is _____ b, denoted _____, if either $a > b$ or $a = b$ is true.

6. In an addition problem the two numbers added are called _____, and the answer is called the _____.

7. If a real number a is located to the left of a real number b on the number line, we say that a is _____ b and we write _____.

Addition of Real Numbers

Exercises 1-10: Refer to Examples 1-3 on pages 45-46 in your text and the Section 1.5 lecture video.

Find the opposite of each number and calculate the sum of the number and its opposite.

1. 32

1. _____

_____

2. $-\sqrt{3}$

2. _____

_____

3. $\dfrac{5}{6}$

3. _____

_____

Find each sum by hand.

4. $-4+(-9)$

4. _____

5. $-\dfrac{3}{4}+\dfrac{4}{5}$

5. _____

6. $5.1+(-9.4)$

6. _____

Add visually, using the symbols $\cap$ and $\cup$.

7. $5+3$

7. _____

8. $-7+4$

8. _____

9. $2+(-6)$

9. _____

10. $-7+(-3)$

10. _____

Subtraction of Real Numbers

Exercises 11-17: Refer to Examples 4-5 on pages 47-48 in your text and the Section 1.5 lecture video.

Find each difference by hand.

11. $8-28$ **11.** _____

12. $-3-9$ **12.** _____

13. $-6.7-(-4.3)$ **13.** _____

14. $-\dfrac{2}{3}-\left(-\dfrac{4}{5}\right)$ **14.** _____

Evaluate each expression by hand.

15. $8-4-(-2)+3$ **15.** _____

16. $-\dfrac{5}{6}+\dfrac{1}{3}-\left(-\dfrac{3}{4}\right)$ **16.** _____

17. $-3.1+6.9-10.2$ **17.** _____

Applications

Exercises 18-19: Refer to Examples 6-7 on page 49 in your text and the Section 1.5 lecture video.

18. In a nine-month period in Punta Arenas, at the southern tip of   **18.** _____
Chile, the temperature, with the wind chill factor, ranged from
$-120°F$ to $52°F$. Find the difference between these temperatures.

19. The initial balance in a checking account is $315. Find the final   **19.** _____
balance if the following represents a list of withdrawals and
deposits: $-$65$, $-$40$, 140, and $-$85$.

Chapter 1 Introduction to Algebra
1.6 Multiplication and Division of Real Numbers

Multiplication of Real Numbers ~ Division of Real Numbers ~ Applications

---

STUDY PLAN

Read: Read Section 1.6 on pages 51-58 in your textbook or eText.

Practice: Do your assigned exercises in your ☐ Book ☐ MyMathLab ☐ Worksheets

Review: Keep your corrected assignments in an organized notebook and use them to review for the test.

Key Terms
Exercises 1-4: Use the vocabulary terms listed below to complete each statement. Note that some terms or expressions may not be used.

quotient   dividend
reciprocal   negative
divisor   multiplicative inverse
positive

1. The product or quotient of two numbers with unlike signs is a(n) _____ number.

2. The _____, or _____, of a real number a is $\dfrac{1}{a}$.

3. The product or quotient of two numbers with like signs is a(n) _____ number.

4. In the division problem $20 \div 4 = 5,$ the number 20 is the _____, 4 is the _____, and 5 is the _____.

Multiplication of Real Numbers

Exercises 1-8: Refer to Examples 1-2 on pages 52-53 in your text and the Section 1.6 lecture video.

Find each product by hand.

1. $-6 \cdot 4$ 1. _____

2. $\dfrac{3}{4} \cdot \dfrac{2}{5}$ 2. _____

3. $-3.2(-1.5)$ 3. _____

4. $(1.2)(-3)(-4)(-5)$ 4. _____

Evaluate each expression by hand.

5. $(-5)^2$ 5. _____

6. -5^2 6. _____

7. $(-4)^3$ 7. _____

8. -4^3 8. _____

Division of Real Numbers

Exercises 9-20: Refer to Examples 3-6 on pages 54-56 in your text and the Section 1.6 lecture video.

Evaluate each expression by hand.

9. $-15 \div \dfrac{1}{3}$

9. _____

10. $\dfrac{\frac{2}{3}}{6}$

10. _____

11. $-\dfrac{-12}{-32}$

11. _____

12. $0 \div (-4)$

12. _____

Convert each measurement to a decimal number.

13. $\dfrac{5}{6}$-inch nail

13. _____

14. $\dfrac{7}{16}$-inch radius

14. _____

15. $1\dfrac{1}{8}$-cup sugar

15. _____

Convert each decimal number to a fraction in lowest terms.

16.   0.08

16. _____

17.   0.275

17. _____

18.   0.005

18. _____

Use a calculator to evaluate each expression. Express your answer as a decimal and as a fraction.

19. $\dfrac{5}{6} - \dfrac{1}{2} + \dfrac{4}{3}$

19. _____

20. $\left(\dfrac{2}{3} \cdot \dfrac{5}{8}\right) \div \dfrac{4}{9}$

20. _____

Applications

Exercises 21-22: Refer to Examples 7-8 on pages 56-57 in your text and the Section 1.6 lecture video.

21.   Lotto winnings of $1.2 million were divided among three people. Because he purchased the ticket and chose the numbers, Sam received $\dfrac{7}{15}$ of the total winnings, while his two friends divided the remaining amount.

(a) What amount did Sam receive?

21.(a)_____

(b) What amount did each of Sam's friends receive?

(b)_____

22.   A student planned to spend $\dfrac{19}{250}$ of his education budget for textbooks. Write this fraction as a decimal.

22. _____

Chapter 1 Introduction to Algebra
1.7 Properties of Real Numbers

Commutative Properties ~ Associative Properties ~ Distributive Properties ~ Identity and Inverse Properties ~ Mental Calculations

---

STUDY PLAN

Read: Read Section 1.7 on pages 60-68 in your textbook or eText.

Practice: Do your assigned exercises in your ☐ Book ☐ MyMathLab ☐ Worksheets

Review: Keep your corrected assignments in an organized notebook and use them to review for the test.

Key Terms
Exercises 1-9: Use the vocabulary terms listed below to complete each statement.
Note that some terms or expressions may not be used.

additive inverse property	commutative property for multiplication
distributive property	identity property of 0
associative property for addition	multiplicative inverse property
identity property of 1	associative property for multiplication
commutative property for addition	

1. The _____ states that for any real number a, $a+0=0+a=a$.

2. The _____ states that for any real numbers a, b and c,
   $(a+b)+c=a+(b+c)$.

3. The _____ states that for any nonzero real number a, $a \cdot \frac{1}{a}=1$ and $\frac{1}{a} \cdot a=1$.

4. The _____ states that for any real numbers a, b and c, $a(b+c)=ab+ac$
   and $a(b-c)=ab-ac$.

5. The _____ states that for any real number, a and b, $a \cdot b=b \cdot a$.

6. The _____ states that if any number a is multiplied by 1, the result is a.

7. The _____ states that for any real numbers a, b and c, $(a \cdot b) \cdot c = a \cdot (b \cdot c)$.

8. The _____ states that for any real number a, $a + (-a) = 0$ and $-a + a = 0$.

9. The _____ states that two numbers a and b can be added in any order and the result will be the same.

Commutative Properties

Exercises 1-2: Refer to Example 1 on page 61 in your text and the Section 1.7 lecture video.

Use a commutative property to rewrite each expression.

1. $4 + 20$ 1. _____

2. $x \cdot 9$ 2. _____

Associative Properties

Exercises 3-7: Refer to Examples 2-3 on page 62 in your text and the Section 1.7 lecture video.

Use an associative property to rewrite each expression.

3. $(2 + 4) + 7$ 3. _____

4. $a(bc)$ 4. _____

State the property that each equation illustrates.

5. $5 + (1 + x) = (5 + 1) + x$ 5. _____

6. $x \cdot y = y \cdot x$ 6. _____

7. $2 + ab = ab + 2$ 7. _____

Distributive Properties

Exercises 8-18: Refer to Examples 4-6 on pages 64-65 in your text and the Section 1.7 lecture video.

Apply a distributive property to each expression.

8. $4(a-5)$ 8. _____

9. $-3(b+8)$ 9. _____

10. $-(x-9)$ 10. _____

11. $14-(y+3)$ 11. _____

Use the distributive property to insert parentheses in the expression and then simplify the result.

12. $7x+4x$ 12. _____

13. $2a-10a$ 13. _____

14. $-6y+3y$ 14. _____

State the property or properties illustrated by each equation.

15. $(3+x)+7=x+10$ 15. _____

16. $4(12a)=48a$ 16. _____

17. $-3(x-9)=-3x+27$ 17. _____

18. $2(b+a)=2a+2b$ 18. _____

Identity and Inverse Properties

Exercises 19-22: Refer to Example 7 on page 66 in your text and the Section 1.7 lecture video.

State the property or properties illustrated by each equation.

19. $\dfrac{12}{32} = \dfrac{3}{8} \cdot \dfrac{4}{4} = \dfrac{3}{8}$

19. _____

20. $-7 + 0 = -7$

20. _____

21. $\dfrac{1}{3} \cdot 3a = 1 \cdot a = a$

21. _____

22. $2 + (-2) + x = 0 + x = x$

22. _____

Mental Calculations

Exercises 23-27: Refer to Examples 8-9 on page 67 in your text and the Section 1.7 lecture video.

Use properties of real numbers to calculate each expression mentally.

23. $\dfrac{4}{3} \cdot 2 \cdot \dfrac{3}{4} \cdot \dfrac{1}{2}$

23. _____

24. $12 + 5 + 8 + 25$

24. _____

25. $3 \cdot 65$

25. _____

26. $434 + 99$

26. _____

27. A tank is 40 feet long, 25 feet wide, and 5 feet deep. The volume V of the tank is found by multiplying 40, 25 and 5. Use the commutative and associative properties for multiplication to calculate the volume of the tank mentally.

27. _____

Chapter 1 Introduction to Algebra
1.8 Simplifying and Writing Algebraic Expressions

Terms ~ Combining Like Terms ~ Simplifying Expressions ~ Writing Expressions

---

STUDY PLAN

Read: Read Section 1.8 on pages 71-76 in your textbook or eText.

Practice: Do your assigned exercises in your ☐ Book ☐ MyMathLab ☐ Worksheets

Review: Keep your corrected assignments in an organized notebook and use them to review for the test.

Key Terms
Exercises 1-5: Use the vocabulary terms listed below to complete each statement.
Note that some terms or expressions may not be used.

    term
    multiplicative identity
    like term
    coefficient
    additive identity

1. The number 0 is called the _____.

2. A(n) _____ is a number, a variable, or a product of numbers and variables raised to natural number powers.

3. The _____ of a term is the number that appears in the term.

4. If two terms contain the same variables raised to the same powers, we call them _____(s).

5. The number 1 is called the _____.

Terms

Exercises 1-4: Refer to Example 1 on page 72 in your text and the Section 1.8 lecture video.

Determine whether each expression is a term. If it is a term, identify its coefficient.

1. -2

    1. _____

2. $4x$

    2. _____

3. $3x - 7y$

    3. _____

4. $6a^3$

    4. _____

Combining Like Terms

Exercises 5-11: Refer to Examples 2-3 on page 73 in your text and the Section 1.8 lecture video.

Determine whether the terms are like or unlike.

5. $-2a, -2b$

    5. _____

6. $4x^2, -9x^2$

    6. _____

7. $3z, 5z^2$

    7. _____

8. $-3m, 8m$

    8. _____

Combine terms in each expression, if possible.

9. $-4x + \dfrac{1}{2}x$

    9. _____

10. $7y^2 - y^2$

    10. _____

11. $-a^2 + 4a$

    11. _____

Simplifying Expressions

Exercises 12-19: Refer to Examples 4-6 on pages 74-75 in your text and the Section 1.8 lecture video.

Simplify each expression.

12. $3 - x - 8 + 4x$

12._____

13. $8y - (y + 13)$

13._____

14. $\dfrac{-2.7a}{-2.7}$

14._____

15. $10 - 3(b + 4)$

15._____

16. $5x^2 - x + 3x^2 + x$

16._____

17. $7t^3 - t^3 - 12t^3$

17._____

18. $\dfrac{12z - 9}{3}$

18._____

19. Simplify the expression $7a + 8 - 2a - 13$.

19._____

Writing Expressions

Exercise 20: Refer to Example 7 on page 76 in your text and the Section 1.8 lecture video.

20. A street has a constant width w and comprises several short sections having lengths 450, 600, 520, and 700 feet.

  (a) Write and simplify an expression that gives the square footage of the street.

20. (a)_____

  (b) Find the area of the street if its width is 48 feet.

  (b)_____

Chapter 2 Linear Equations and Inequalities
2.1 Introduction to Equations

Basic Concepts ~ Equations and Solutions ~ The Addition Property of Equality ~ The Multiplication Property of Equality

---

STUDY PLAN

Read: Read Section 2.1 on pages 90-97 in your textbook or eText.

Practice: Do your assigned exercises in your   ☐ Book   ☐ MyMathLab   ☐ Worksheets

Review: Keep your corrected assignments in an organized notebook and use them to review for the test.

Key Terms
Exercises 1-5: Use the vocabulary terms listed below to complete each statement.
Note that some terms or expressions may not be used.

> **solution**
> **equivalent**
> **solution set**
> **addition property of equality**
> **multiplication property of equality**

1. The _____ states that if a, b, and c are real numbers, then $a = b$ is equivalent to $a + c = b + c$.

2. The set of all solutions to an equation is called the _____.

3. The _____ states that multiplying each side of an equation by the same nonzero number results in an equivalent equation.

4. _____ equations are equations that have the same solution set.

5. Each value of the variable that makes an equation true is called a(n) _____ to the equation.

The Addition Property of Equality

Exercises 1-4: Refer to Examples 1-2 on pages 92-93 in your text and the Section 2.1 lecture video.

Solve each equation.

1. $x + 12 = 6$ 1. _____

2. $t - 2 = 9$ 2. _____

3. $\dfrac{2}{3} = -\dfrac{1}{2} + x$ 3. _____

4. Solve the equation $-6 + y = -5$ and then check the solution.   4. _____

The Multiplication Property of Equality

Exercises 5-9: Refer to Examples 3-5 on pages 94-96 in your text and the Section 2.1 lecture video.

Solve each equation.

5. $\dfrac{1}{3}x = -3$ 5. _____

6. $-3t = 15$ 6. _____

7. $6 = \dfrac{2}{3}x$ 7. _____

8. Solve the equation $\dfrac{3}{4} = -\dfrac{3}{2}z$ and then check the solution.   8. _____

9. A 2009 study concluded that 24 cubic miles of Alaskan glacier ice are lost each year.

    (a) Write a formula that gives the number of cubic miles of   9. (a)_____
glacier ice lost in x years.

    (b) At this rate, how many years will it take for 600 cubic   (b)_____
miles of glacier ice to melt?

Chapter 2 Linear Equations and Inequalities
2.2 Linear Equations

Basic Concepts ~ Solving Linear Equations ~ Applying the Distributive Property ~ Clearing Fractions and Decimals ~ Equations with No Solutions or Infinitely Many Solutions

---

STUDY PLAN

Read: Read Section 2.2 on pages 99-108 in your textbook or eText.

Practice: Do your assigned exercises in your   ☐ Book   ☐ MyMathLab   ☐ Worksheets

Review: Keep your corrected assignments in an organized notebook and use them to review for the test.

Key Terms
Exercises 1-6: Use the vocabulary terms listed below to complete each statement. Note that some terms or expressions may not be used.

identity	**linear equation**
no solutions	**contradiction**
one solution	**infinitely many solutions**

1. A(n) _____ in one variable is an equation that can be written in the form $ax + b = 0$, where a and b are constants with $a \neq 0$.

2. If the process of solving an equation results in a contradiction, such as $0 = 4$ or $5 = 1$, the equation has _____ .

3. If the process of solving an equation results in an identity, such as $0 = 0$ or $-2 = -2$, the equation has _____ .

4. An equation that is always false is called a(n) _____ .

5. If the process of solving an equation results in a statement which is true for only one value, such as $x = 3$ or $x = -7$, the equation has _____ .

6. An equation that is always true is called a(n) _____ .

Basic Concepts

Exercises 1-4: Refer to Example 1 on pages 100-101 in your text and the Section 2.2 lecture video.

Determine whether the equation is linear. If the equation is linear, give values for a and b that result when the equation is written in the form $ax + b = 0.$

1. $3x - 2 = 0$ 1. _____

2. $4x^2 - 3 = 0$ 2. _____

3. $\dfrac{1}{2} = 7$ 3. _____

4. $\dfrac{2}{3}x - 6 = 0$ 4. _____

Solving Linear Equations

Exercises 5-9: Refer to Examples 2-4 on pages 101-104 in your text and the Section 2.2 lecture video.

5. Complete the table for the given values of x.   5. _____
    Then solve the equation $-2x - 5 = -1.$

x	-3	-2	-1	0	1	2	3
$-2x - 5$							

Solve each linear equation. Check the answer.

6. $4x - 7 = 0$ 6. _____

7. $\dfrac{1}{3}x + 4 = 2$ 7. _____

8. $6x - 1 = 2x + 9$ 8. _____

9. The number of Internet users I in millions during year x can be approximated by the formula $I = 241x - 482,440$, where $x \geq 2007$. Estimate the year when there were 1970 million (1.97 billion) Internet users.

9. _____

Applying the Distributive Property

Exercises 10-11: Refer to Example 5 on pages 104-105 in your text and the Section 2.2 lecture video.

Solve each linear equation. Check the answer.

10. $3(x+5)+x=0$

10. _____

11. $3(2z-7)+4=4(z-1)$

11. _____

Clearing Fractions and Decimals

Exercises 12-15: Refer to Examples 6-7 on pages 105-106 in your text and the Section 2.2 lecture video.

Solve each linear equation.

12. $\dfrac{4}{3} = \dfrac{1}{6} - \dfrac{5}{4}b$

12. _____

13. $\dfrac{2}{3}x - \dfrac{3}{4} = \dfrac{1}{2} - x$

13. _____

14. $2.5x - 1.4 = 4.2$

14. _____

15. $5.1 - 2.2y = y + 3.7$

15. _____

Equations with No Solutions or Infinitely Many Solutions

Exercises 16-18: Refer to Example 8 on page 107 in your text and the Section 2.2 lecture video.

Determine whether the equation has no solutions, one solution, or infinitely many solutions.

16. $3x - 2(x - 2) = x + 4$

16. _____

17. $3x = 5x + 2(x + 4)$

17. _____

18. $6x + 7 = 2(3x - 2) + 2$

18. _____

Chapter 2 Linear Equations and Inequalities
2.3 Introduction to Problem Solving

Steps for Solving a Problem ~ Percent Problems ~ Distance Problems ~ Other Types of Problems

---

STUDY PLAN

Read: Read Section 2.3 on pages 111-120 in your textbook or eText.

Practice: Do your assigned exercises in your ☐ Book ☐ MyMathLab ☐ Worksheets

Review: Keep your corrected assignments in an organized notebook and use them to review for the test.

Key Terms
Exercises 1-7: Use the vocabulary terms listed below to complete each statement.
Note that some terms or expressions may not be used.

is	less	per
plus	double	add
subtract	total	sum
decimal number	results in	gives
quotient	fraction	minus
equals	product	increase
decrease	divide	fewer
multiply	times	triple
more	difference	percent change
twice	divided by	is the same as

1. If a quantity changes from an old amount to a new amount , the _____ is given by $\dfrac{\text{new amount} - \text{old amount}}{\text{old amount}} \cdot 100$.

2. The expression $x\%$ represents the _____ $\dfrac{x}{100}$ or the _____ $x \cdot 0.01$.

3. List six words/phrases that are associated with the math symbol $+$.

4. List six words/phrases that are associated with the math symbol $-$.

5. List six words/phrases that are associated with the math symbol $\cdot$.

6. List four words/phrases that are associated with the math symbol $\div$.

7. List five words/phrases that are associated with the math symbol $=$.

Steps for Solving a Problem

Exercises 1-5: Refer to Examples 1-3 on pages 112-113 in your text and the Section 2.3 lecture video.

Translate the sentence into an equation using the variable x. Then solve the resulting equation.

1. Four times a number plus 5 is equal to 17.

        1. _____

2. The sum of one-third a number and 7 is 1.

        2. _____

3. Fifteen is 3 less than twice a number.

        3. _____

4. The sum of three consecutive natural numbers is 54.
    Find the three numbers.

        4. _____

5. The population of a small town is 34,000. This is 2000 more
    than twice the population of a neighboring town. Find the
    population of the neighboring town.

        5. _____

Percent Problems

Exercises 6-14: Refer to Examples 4-8 on pages 114-116 in your text and the Section 2.3 lecture video.

Convert each percentage to fraction and decimal notation.

6. 33%

        6. _____

7. 2.5%

        7. _____

8. 0.8%

        8. _____

Convert each real number to a percentage.

9. 0.137

9. _____

10. $\dfrac{1}{8}$

10. _____

11. 1.61

11. _____

12. From 2005 to 2010, the cost of tuition increased from $130 to $156 per credit. Calculate the percent increase in tuition cost from 2005 to 2010.

12. _____

13. In 2010, an office manager received a salary of $57,500. With the downturn in the economy in 2011, he was forced to take a 3% pay cut. Calculate his salary after the cut.

13. _____

14. In 2050, about 16% of the world's population, or 1.5 billion people, will be older than 65. Find the estimated population of the world in 2050.

14. _____

Distance Problems

Exercises 15-16: Refer to Examples 9-10 on pages 116-117 in your text and the Section 2.3 lecture video.

15.   A pilot flies a plane at a constant speed for 3 hours and   15. _____
      20 minutes, traveling 1900 miles. Find the speed of the
      plane in miles per hour.

16.   A marathoner jogs at two speeds, covering a distance of   16. _____
      15 miles in 2 hours. If she runs one-half hour at 6 miles
      per hour, find the second speed.

Other Types of Problems

Exercises 17-18: Refer to Examples 11-12 on pages 118-119 in your text and the Section 2.3 lecture video.

17.   A solution contains 5% salt. How much pure water should   17. _____
      be added to 20 ounces of the solution to dilute it to a 2%
      solution?

18.   A student takes out a loan for a limited amount of money at   18. _____
      9% interest and then must pay 11% for any additional money.
      If the student borrows $3500 less at 11% than at 9%, then the
      total interest for one year is $615. How much does the student
      borrow at each rate?

Chapter 2 Linear Equations and Inequalities
2.4 Formulas

Formulas from Geometry ~ Solving for a Variable ~ Other Formulas

---

STUDY PLAN

Read: Read Section 2.4 on pages 124-133 in your textbook or eText.

Practice: Do your assigned exercises in your ☐ Book ☐ MyMathLab ☐ Worksheets

Review: Keep your corrected assignments in an organized notebook and use them to review for the test.

Key Terms
Exercises 1-8: Use the vocabulary terms listed below to complete each statement. Note that some terms or expressions may not be used. Some terms may be used more than once.

> **perimeter**
> **area**
> **degree**
> **circumference**
> **volume**

1. The _____ of a circle with radius r is given by πr^2.

2. If a rectangle has length l and width w, then its _____ is given by lw.

3. The _____ of a cylinder having radius r and height h is given by $\pi r^2 h$.

4. The perimeter of a circle is called its _____.

5. If a rectangle has length l and width w, then its _____ is given by $2l + 2w$.

6. A(n) _____ is $\dfrac{1}{360}$ of a revolution.

7. If a triangle has base b and height h, then its _____ is given by $\dfrac{1}{2}bh$.

8. The _____ of a circle with radius r is given by $2\pi r$.

Formulas from Geometry

Exercises 1-6: Refer to Examples 1-6 on pages 125-128 in your text and the Section 2.4 lecture video.

1. A residential lot is shown in the figure. It comprises a rectangular region and an adjacent triangular region.

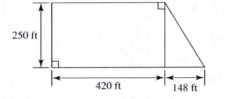

    (a) Find the area of this lot.

    1.(a)_____

    (b) An acre contains 43,560 square feet. How many acres are there in this lot?

    (b)_____

2. In a triangle, the two larger angles are equal in measure and each is twice the measure of the smallest angle. Find the measure of each angle.

    2. _____

    _____

    _____

3. A circle has a diameter of 15 centimeters. Find its circumference and area.

    3. _____

    _____

4. Find the area of the trapezoid shown in the figure.

    4. _____

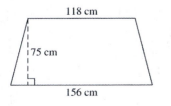

5. Find the volume and surface area of a box with length 12 meters, width 5 meters, and height 2 meters.

5. _____

_____

6. A cylindrical soup can has a radius $1\frac{1}{4}$ of inches and a height of 4 inches.

(a) Find the volume of the can.

6.(a)_____

(b) If 1 cubic inch equals 0.554 fluid ounce, find the number of fluid ounces in the can.

(b)_____

Solving for a Variable

Exercises 7-9: Refer to Examples 7-8 on pages 129-130 in your text and the Section 2.4 lecture video.

7. The perimeter of a rectangle is given by $P = 2l + 2w$.

(a) Solve the formula for l.

7.(a)_____

(b) A rectangle has perimeter $P = 26$ inches and width $w = 5$ inches. Find l.

(b)_____

Solve each equation for the indicated variable.

8. $A = \frac{1}{2}(a+b)h$ for h

8. _____

9. $xy + yz = xz$ for x

9. _____

Other Formulas

Exercises 10-11: Refer to Examples 9-10 on pages 130-131 in your text and the Section 2.4 lecture video.

10.   A student has earned 12 credits of A, 30 credits of B, 22 credits   10. _____
      of C, 4 credits of D, and 4 credits of F. Calculate the student's
      GPA to the nearest hundredth.

11.   To convert Celsius degrees C to Fahrenheit degrees F, the formula
      $F = \dfrac{9}{5}C + 32$ can be used.

      (a) Solve the formula for C to find a formula that converts   **11. (a)**_____
      Fahrenheit degrees to Celsius degrees.

      (b) If the temperature is 77°F, find the equivalent Celsius   **(b)**_____
      temperature.

Chapter 2 Linear Equations and Inequalities
2.5 Linear Inequalities

Solutions and Number Line Graphs ~ The Addition Property of Inequalities ~ The Multiplication Property of Inequalities ~ Applications

---

STUDY PLAN

Read: Read Section 2.5 on pages 136-145 in your textbook or eText.

Practice: Do your assigned exercises in your ☐ Book ☐ MyMathLab ☐ Worksheets

Review: Keep your corrected assignments in an organized notebook and use them to review for the test.

Key Terms
Exercises 1-5: Use the vocabulary terms listed below to complete each statement.
Note that some terms or expressions may not be used.

> **solution**
> **solution set**
> **interval notation**
> **linear inequality**
> **set-builder notation**

1. A solution in the form $(-3, 4]$ is written in _____.

2. The set of all solutions to an inequality is called the _____.

3. A solution in the form $\{x \mid x \le -2\}$ is written in _____.

4. A(n) _____ results whenever the equals sign in a linear equation is replaced with any one of the symbols $<, \le, >,$ or $\ge$.

5. A(n) _____ to an inequality is a value of the variable that makes the statement true.

Solutions and Number Line Graphs

Exercises 1-13: Refer to Examples 1-4 on pages 137-138 in your text and the Section 2.5 lecture video.

Use a number line to graph the solution set to each inequality.

1. $x < 0$

2. $x \leq 0$

3. $x > 2$

4. $x \leq 5$

Write the solution set to each inequality in interval notation.

5. $x < 7$ 5._____

6. $y \geq -4$ 6._____

7. $a > 0$ 7._____

Determine whether the given value of x is a solution to the inequality.

8. $2x - 7 \leq 5, \ x = 3$ 8._____

9. $6 - 4x > 10, \ x = -1$ 9._____

In the table, the expression $-3x+5$ has been evaluated for several values of x.
Use the table to determine any solutions to each equation or inequality.

x	-2	-1	0	1	2	3	4
$-3x+5$	11	8	5	2	-1	-4	-7

10. $-3x+5=5$

10. _____

11. $-3x+5>5$

11. _____

12. $-3x+5\geq5$

12. _____

13. $-3x+5<5$

13. _____

The Addition Property of Inequalities

Exercises 14-16: Refer to Examples 5-6 on pages 139-140 in your text and the Section 2.5 lecture video.

Solve each inequality. Then graph the solution set.

14. $x+3\leq-6$

14. _____

15. $4+2x>7+x$

15. _____

16. $6-\dfrac{1}{2}x\geq4+\dfrac{1}{2}x$

16. _____

The Multiplication Property of Inequalities

Exercises 17-21: Refer to Examples 7-8 on pages 141-143 in your text and the Section 2.5 lecture video.

Solve each inequality. Then graph the solution set.

17. $-2x \leq 14$ 17. _____

18. $-3 > -\dfrac{1}{3}x$ 18. _____

Solve each inequality. Write the solution set in set-builder notation.

19. $3x - 8 < -6$ 19. _____

20. $-6 + 6x \leq 5x + 2$ 20. _____

21. $0.3(2x - 1) > -2.3x - 9$ 21. _____

Applications

Exercises 22-26: Refer to Examples 9-11 on pages 143-144 in your text and the Section 2.5 lecture video.

Translate each phrase to an inequality. Let the variable be x.

22. A number that is less than −10

22. _____

23. A temperature that is at most 90°F

23. _____

24. A grade point average that is at least 3.5

24. _____

25. If the ground temperature is 81°F, then the air temperature x miles high is given by the formula $T = 81 - 19x$. Determine the altitudes at which the air temperature is less than 5°F.

25. _____

26. To manufacture a certain product, a company incurs a one-time fixed cost of \$4200 plus a per-item cost of \$180. The selling price of the item is \$240.

(a) Write a formula that gives the cost C of producing x items.

26. (a)_____

(b) Write a formula that gives the revenue R from selling x items.

(b)_____

(c) Profit equals revenue minus cost. Write a formula that calculates the profit P from producing and selling x items.

(c)_____

(d) How many items must be sold to yield a positive profit?

(d)_____

Chapter 3 Graphing Equations
3.1 Introduction to Graphing

Tables and Graphs ~ The Rectangular Coordinate System ~ Scatterplots and Line Graphs

---

STUDY PLAN

Read: Read Section 3.1 on pages 159-165 in your textbook or eText.

Practice: Do your assigned exercises in your   ☐ Book   ☐ MyMathLab   ☐ Worksheets

Review: Keep your corrected assignments in an organized notebook and use them to review for the test.

Key Terms
Exercises 1-8: Use the vocabulary terms listed below to complete each statement.
Note that some terms or expressions may not be used.

y-axis	**scatterplot**
origin	**y-coordinate**
x-axis	**ordered pair**
line graph	**x-coordinate**
quadrants	**rectangular coordinate system**

1.   In the xy-plane, a point is graphed as a(n) _____ (x, y).

2.   If consecutive data points are connected with line segments, then the resulting graph is called a(n) _____.

3.   When plotting in the xy-plane, the first value in an ordered pair is called the _____ and the second value is called the _____.

4.   The point of intersection of the axes is called the _____ and is associated with zero on each axis.

5.   If distinct points are plotted in the xy-plane, then the resulting graph is called a(n) _____.

6.   The axes divide the xy-plane into four regions called _____, which are numbered I, II, III, and IV counterclockwise.

7.   One common way to graph data is to use the _____, or xy-plane.

8.   In the xy-plane the horizontal axis is the _____, and the vertical axis is the _____.

The Rectangular Coordinate System

Exercises 1-4: Refer to Examples 1-2 on pages 160-161 in your text and the Section 3.1 lecture video.

Plot the following ordered pairs on the same xy-plane. State the quadrant in which each point is located, if possible.

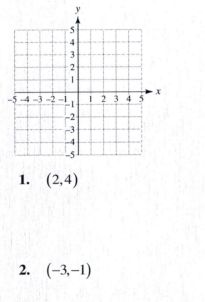

1. $(2, 4)$ 1. _____

2. $(-3, -1)$ 2. _____

3. $(0, -2)$ 3. _____

4. The figure shows the average number of hours of video posted   4. _____
    to YouTube every minute during selected months. Use the graph
    to estimate the number of hours of video posted to YouTube every
    minute in December 2008 and June 2010.

Hours of Video Posted to YouTube
Every Minute

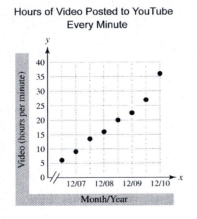

Scatterplots and Line Graphs

Exercises 5-7: Refer to Examples 3-5 on pages 162-164 in your text and the Section 3.1 lecture video.

5.   The table lists the average price of a gallon of milk for selected years. Make a scatterplot of the data.

Year	2006	2007	2008	2009	2010	2011
Cost (per gal)	$2.99	$3.45	$2.65	$2.69	$2.79	$3.39

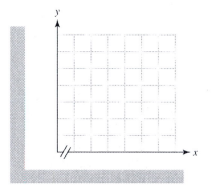

6.   Use the data in the table to make a line graph.   6. _____

x	−4	−2	0	2	4
y	5	3	−4	−2	1

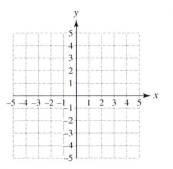

7. The line graph in the figure shows the per capita sugar consumption in the United States from 1950 to 2000.

7. _____

Sugar Consumption in the United States

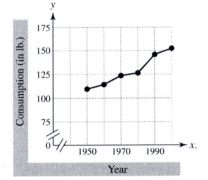

(a) Did sugar consumption ever decrease during this time period?

7. (a)_____

(b) Estimate the sugar consumption in 1960 and in 2000.

(b)_____

(c) Estimate the percent change in sugar consumption from 1960 to 2000.

(c)_____

Chapter 3 Graphing Equations
3.2 Linear Equations

Basic Concepts ~ Table of Solutions ~ Graphing Linear Equations in Two Variables

---

STUDY PLAN

Read: Read Section 3.2 on pages 168-175 in your textbook or eText.

Practice: Do your assigned exercises in your ☐ Book ☐ MyMathLab ☐ Worksheets

Review: Keep your corrected assignments in an organized notebook and use them to review for the test.

Key Terms
Exercises 1-5: Use the vocabulary terms listed below to complete each statement. Note that some terms or expressions may not be used.

one	no
solution	ordered pair
standard form	infinitely many
linear equation in two variables	table

1. The graph of a(n) _____ is a line.

2. Equations in two variables often have _____ solution(s).

3. A(n) _____ can be used to list solutions to an equation in two variables.

4. An ordered pair (x, y) whose x- and y- values satisfy the equation is called a(n) _____ to an equation in two variables.

5. The _____ of a linear equation in two variables is $Ax + By = C$, where A, B, and C are fixed numbers (constants) and A and B are not both equal to 0.

Basic Concepts

Exercises 1-3: Refer to Example 1 on page 169 in your text and the Section 3.2 lecture video.

Determine whether the given ordered pair is a solution to the given equation.

1. $y = x - 4,\ (-3, 1)$ 1. _____

2. $2x + y = 4;\ \left(-\dfrac{1}{2}, 3\right)$ 2. _____

3. $-3x + 4y = 11;\ (-1, 2)$ 3. _____

Tables of Solutions

Exercises 4-6: Refer to Examples 2-4 on pages 170-171 in your text and the Section 3.2 lecture video.

4.   Complete the table for the equation $y = 3x + 4$.   4. _____

x	-4	-2	0	2
y				

5.   Use $y = 0, 4, 8,$ and 12 to make a table of solutions to $4x + 3y = 12$.   5. _____

x				
y	0	4	8	12

6.   The Asian-American population P in millions t years after the
year 2000 is estimated by $P = 0.4t + 11.2$.

 (a) Complete the table.

t	2	4	6	8	10
P					

 (b) Use the table to determine the Asian-American population   6.(b)_____
 in the year 2008.

Graphing Linear Equations in Two Variables

Exercises 7-10: Refer to Examples 5-7 on pages 171-174 in your text and the Section 3.2 lecture video.

7. Make a table of values for the equation $y = -3x$, and then   7. _____
    use the table to graph this equation.

x	-2	-1	0	1
y				

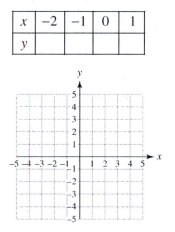

Graph each linear equation.

8. $y = -\dfrac{1}{2}x + 3$

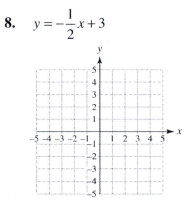

9. $x + y = -2$

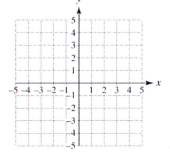

10. Graph the linear equation $3x - 5y = 15$ by solving for y first.

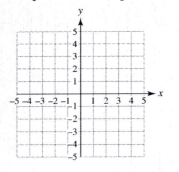

Name: _____ **Course/Section:** _____ **Instructor:** _____

Chapter 3 Graphing Equations
3.3 More Graphing of Lines

Finding Intercepts ~ Horizontal Lines ~ Vertical Lines

STUDY PLAN

Read: Read Section 3.3 on pages 178-186 in your textbook or eText.

Practice: Do your assigned exercises in your ☐ Book ☐ MyMathLab ☐ Worksheets

Review: Keep your corrected assignments in an organized notebook and use them to review for the test.

Key Terms

Exercises 1-4: Use the vocabulary terms listed below to complete each statement. Note that some terms or expressions may not be used.

x	y
$x = 0$	$y = 0$
x-axis	y-axis
x-coordinate	y-coordinate
vertical line	horizontal line

1. The equation of a _____ with y-intercept b is $y = b$.

2. The _____ of a point where a graph intersects the _____ is called a y-intercept. To find a y-intercept, let _____ in the equation and solve for _____.

3. The equation of a _____ with x-intercept k is $x = k$.

4. The _____ of a point where a graph intersects the _____ is called an x-intercept. To find the x-intercept, let _____ in the equation and solve for _____.

Finding Intercepts

Exercises 1-3: Refer to Examples 1-3 on pages 180-181 in your text and the Section 3.3 lecture video.

1. Use intercepts to graph $-5x + 2y = 10$.

    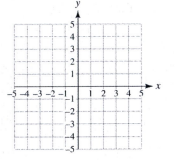

    1. _____

2. Complete the table for the equation $x - y = -2$. Then determine the x-intercept and y-intercept for the graph of the equation $x - y = -2$.

x	-2	-1	0	1	2
y					

    2. _____

    _____

3. A ball is thrown into the air. Its velocity v in feet per second after t seconds is given by $v = 80 - 32t$. Assume $t \geq 0$ and $t \leq 2.5$.

    (a) Graph the equation by finding the intercepts. Let t (time) correspond to the horizontal axis (x-axis) and v (velocity) correspond to the vertical axis (y-axis).

    3. **(a)**

    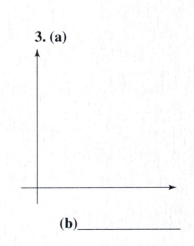

    (b) Interpret each intercept.

    (b)_____

Horizontal Lines

Exercise 4: Refer to Example 4 on page 182 in your text and the Section 3.3 lecture video.

4. Graph the equation $y = -3$ and identify its y-intercept.   4. _____

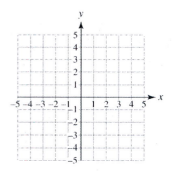

Vertical Lines

Exercise 5-10: Refer to Examples 5-7 on page 184-185 in your text and the Section 3.3 lecture video.

5. Graph the equation $x = -1$ and identify its x-intercept.   5. _____

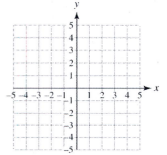

Write the equation of the line shown in each graph.

6.   6. _____

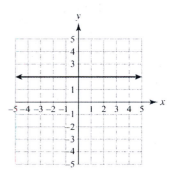

7.

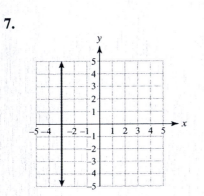

7. _____

Find an equation for a line satisfying the given conditions.

8. Vertical, passing through $(-4, 4)$

8. _____

9. Horizontal, passing through $(5, 1)$

9. _____

10. Perpendicular to $y = -2$, passing through $(0, -6)$

10. _____

Chapter 3 Graphing Equations
3.4 Slope and Rates of Change

Finding Slopes of Lines ~ Slope as a Rate of Change

STUDY PLAN

Read: Read Section 3.4 on pages 190-198 in your textbook or eText.

Practice: Do your assigned exercises in your ☐ Book ☐ MyMathLab ☐ Worksheets

Review: Keep your corrected assignments in an organized notebook and use them to review for the test.

Key Terms
Exercises 1-8: Use the vocabulary terms listed below to complete each statement.
Note that some terms or expressions may not be used.

rise	**run**
slope	**negative slope**
positive slope	**rate of change**
zero slope	**undefined slope**

1. A line with _____ is horizontal.

2. If a line has _____, it falls from left to right.

3. The _____, or change in x, is $x_2 - x_1$.

4. The _____ m of the line passing through the points (x_1, y_1) and (x_2, y_2) is
    $m = \dfrac{y_2 - y_1}{x_2 - x_1}$, where $x_1 \neq x_2$.

5. Slope measures the _____ in a quantity.

6. The _____, or change in y, is $y_2 - y_1$.

7. If a line has _____, it rises from left to right.

8. A vertical line has _____.

Finding Slopes of Lines

Exercises 1-9: Refer to Examples 1-6 on pages 191-194 in your text and the Section 3.4 lecture video.

1. Use the two points labeled in the figure to find the slope of the line.   1. _____
    What are the rise and run between these two points? Interpret the
    slope in terms of rise and run.

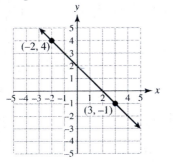

Calculate the slope of the line passing through each pair of points, if possible. Graph the line.

2. $(-4,1)$ and $(2,1)$ 2. _____

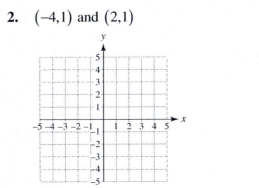

3. $(2,-2)$ and $(5,4)$ 3. _____

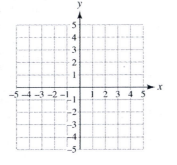

4. $(-3,4)$ and $(-3,-1)$ **4.** _____

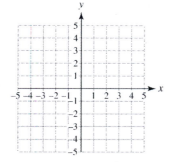

Find the slope of each line.

5.   **5.** _____

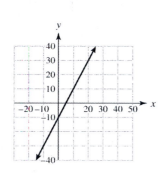

6.   **6.** _____

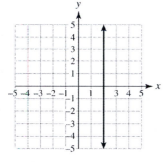

7. Sketch a line passing through the point $(-2,3)$ and having slope $-\dfrac{1}{4}$.   **7.** _____

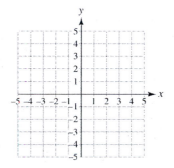

8. Sketch a line with slope −4 and y-intercept 1.

8. _____

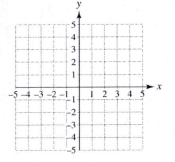

9. A line has slope −3 and passes through the first point listed in the table. Complete the table so that each point lies on the line.

9. _____

x	−2	−1	0	1
y	4			

Slope as a Rate of Change

Exercises 10-13: Refer to Examples 7-10 on pages 195-197 in your text and the Section 3.4 lecture video.

10.   The distance y in miles that a cyclist training for a century ride is from home after x hours is shown in the figure.

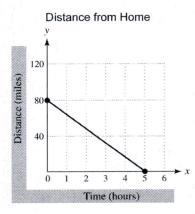

Distance from Home

(a) Find the y-intercept. What does it represent?

10. (a)_____

(b) The graph passes through the point (2, 48). Discuss the meaning of this point.

(b)_____

(c) Find the slope of this line. Interpret the slope as a rate of change.

(c)_____

11. When a company manufactures 1000 game consoles,
its profit is $20,000, and when it manufactures 1500
game consoles, its profit is $35,000.

    (a) Find the slope of the line passing through $(1000, 20{,}000)$ 11. (a)_____

       and $(1500, 35{,}000)$.

    (b) Interpret the slope as a rate of change.   (b)_____

12. The table lists the number of Baccalaureate degrees awarded after
a private two-year college became a four-year institution.

Year	2002	2003	2005	2007	2009	2011
Baccalaureate Degrees	84	706	2206	2443	2726	3057

    (a) Make a line graph of the data.   12. (a)_____

    (b) Find the slope of each line segment.   (b)_____

    (c) Interpret each slope as a rate of change.   (c)_____

13. During a storm, snow falls at the constant rates of 3 inches
per hour from 2 P.M. to 5 P.M., 2 inches per hour from 5 P.M.
to 7 P.M. and $\frac{1}{2}$ per hour from 7 P.M. to 9 P.M.

    (a) Sketch a graph that shows the total accumulation of snowfall   13. (a)_____
       from 2 P.M. to 9 P.M.

    (b) What does the slope of each line segment represent?   (b)_____

Chapter 3 Graphing Equations
3.5 Slope-Intercept Form

Basic Concepts ~ Finding Slope-Intercept Form ~ Parallel and Perpendicular Lines

---

STUDY PLAN

Read: Read Section 3.5 on pages 204-211 in your textbook or eText.

Practice: Do your assigned exercises in your ☐ Book ☐ MyMathLab ☐ Worksheets

Review: Keep your corrected assignments in an organized notebook and use them to review for the test.

Key Terms
Exercises 1-7: Use the vocabulary terms listed below to complete each statement.
Note that some terms or expressions may not be used.

parallel	slope
point-slope	zero slope
perpendicular	slope-intercept
undefined slope	negative reciprocal

1. A line with _____ is horizontal.

2. Two nonvertical _____ lines have the same slope.

3. The _____ m of the line passing through the points (x_1, y_1) and (x_2, y_2) is
   $m = \dfrac{y_2 - y_1}{x_2 - x_1}$, where $x_1 \neq x_2$.

4. If two lines have slopes m_1 and m_2 such that $m_1 \cdot m_2 = -1$, then they are _____ lines.

5. The _____ form of a line with slope m and y-intercept b is given by $y = mx + b$.

6. The slopes of two perpendicular lines are _____(s) of each other.

7. A vertical line has _____.

Finding Slope-Intercept Form

Exercises 1-7: Refer to Examples 1-5 on pages 205-208 in your text and the Section 3.5 lecture video.

For each graph write the slope-intercept form of the line.

1.

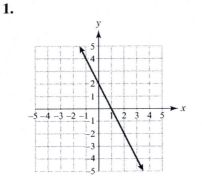

1. _____

2.

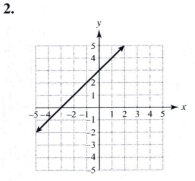

2. _____

3. Sketch a line with slope $\frac{1}{3}$ and y-intercept -4. Write its slope-intercept form.

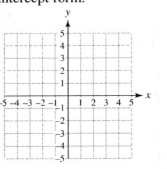

3. _____

Write each equation in slope-intercept form. Then give the slope and y-intercept of the line.

4. $3y - 5x = 15$ 4. _____

_____

_____

5. $x = -3y + 6$ 5. _____

_____

_____

6. Write the equation $3x - y = 2$ in slope-intercept form and   6. _____
   then graph it.

7. Production of a certain item involves fixed costs of $34,000 plus
   $120 for each item made.

   (a) How much does it cost to produce 2000 items?   7. (a)_____

   (b) Write the slope-intercept form that gives the cost to produce   (b)_____
       x items.

   (c) If the cost is $454,000, how many items were produced?   (c)_____

Parallel and Perpendicular Lines

Exercises 8-12: Refer to Examples 6-8 on pages 208-210 in your text and the Section 3.5 lecture video.

8. Find the slope-intercept form of a line parallel to $y = 2x - 7$ 8. _____

    and passing through the point $(-2,1)$. Sketch each line in the

    same xy-plane.

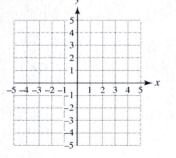

For each of the given lines, find the slope-intercept form of a line passing through the origin that is perpendicular to the given line.

9. $y = -4x$ 9. _____

10. $y = \dfrac{2}{3}x - 4$ 10. _____

11. $5x + 2y = -10$ 11. _____

12. Find the slope-intercept form of a line perpendicular to $y = \dfrac{3}{4}x + 2$ 12. _____

    and passing through the point $(3,-1)$. Sketch each line in the same

    xy-plane.

Chapter 3 Graphing Equations
3.6 Point-Slope Form

Derivation of Point-Slope Form ~ Finding Point-Slope Form ~ Applications

---

STUDY PLAN

   Read: Read Section 3.6 on pages 214-221 in your textbook or eText.

   Practice: Do your assigned exercises in your   ☐ Book ☐ MyMathLab ☐ Worksheets

   Review: Keep your corrected assignments in an organized notebook and use them to review for the test.

Key Terms
Exercises 1-3: Use the vocabulary terms listed below to complete each statement.
Note that some terms or expressions may not be used.

   slope
   point-slope
   slope-intercept

1. The line with slope m passing through the point (x_1, y_1) is given by $y - y_1 = m(x - x_1)$, or equivalently, $y = m(x - x_1) + y_1$. This is called the _____ form of a line.

2. The _____ m of the line passing through the points (x_1, y_1) and (x_2, y_2) is $m = \dfrac{y_2 - y_1}{x_2 - x_1}$, where $x_1 \neq x_2$.

3. The _____ form of a line with slope m and y-intercept b is given by $y = mx + b$.

Derivation of Point-Slope Form

Writing Exercise: Use the slope formula to derive the point-slope form of a line.

Finding Point-Slope Form

Exercises 1-8: Refer to Examples 1-5 on pages 215-219 in your text and the Section 3.6 lecture video.

Use the labeled point in each figure to write a point-slope form for the line and then simplify it to the slope-intercept form.

1.

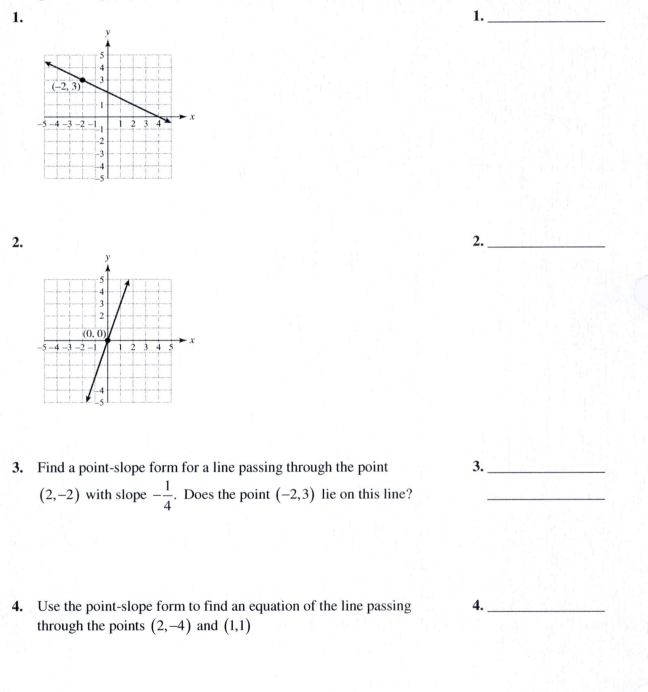

1. _____

2.

2. _____

3. Find a point-slope form for a line passing through the point

$(2,-2)$ with slope $-\dfrac{1}{4}$. Does the point $(-2,3)$ lie on this line?

3. _____

_____

4. Use the point-slope form to find an equation of the line passing through the points $(2,-4)$ and $(1,1)$

4. _____

Find the slope-intercept form for the line that satisfies the given conditions.

5. Slope $-\dfrac{1}{2}$, passing through $(4,-1)$

    5. _____

6. x-intercept -4, y-intercept 1

    6. _____

7. Perpendicular to $3x+4y=5$, passing through $\left(\dfrac{3}{4},-2\right)$

    7. _____

8. The points in the table lie on a line. Find the slope-intercept form of the line.

x	-2	-1	0	1
y	7	4	1	-2

    8. _____

Applications

Exercises 9-10: Refer to Examples 6-7 on pages 219-220 in your text and the Section 3.6 lecture video.

9. In 2005, a self-employed man earned \$56,000. In 2011, he earned \$74,000.

    (a) Find a point-slope form of the line passing through the points $(2006, 56)$ and $(2011, 74)$.

    9. (a)_____

    (b) Interpret the slope as a rate of change.

    (b)_____

    (c) Estimate the man's salary in the year 2014.

    (c)_____

10.   A tank is being emptied by a pump that removes water at a
constant rate. After 2 hours, the tank contains 4500 gallons
of water, and after 5 hours, the tank contains 2250 gallons
of water.

(a) How fast is the pump removing water?   10.(a)_____

(b) Find the slope-intercept form of a line that models the   (b)_____
amount of water in the tank. Interpret the slope.

(c) Find the y-intercept and the x-intercept. Interpret each.   (c)_____

(d) Sketch a graph of the amount of water in the tank   (d)_____
during the first 8 hours.

(e) The point $(4, 3000)$ lies on the graph. Explain its   (e)_____
meaning.

Chapter 3 Graphing Equations
3.7 Introduction to Modeling

Basic Concepts ~ Modeling Linear Data

STUDY PLAN

   Read: Read Section 3.7 on pages 225-229 in your textbook or eText.

   Practice: Do your assigned exercises in your   ☐ Book   ☐ MyMathLab   ☐ Worksheets

   Review: Keep your corrected assignments in an organized notebook and use them to review for the test.

Basic Concepts

Exercise 1: Refer to Example 1 on page 226 in your text and the Section 3.7 lecture video.

1.   After driving 100 miles, a traveler begins to track his mileage. The table below shows the cumulative distance driven after selected days. Does the equation $D = 85x + 100$ model the data exactly? Explain.

1. _____

x	3	6	7
D	355	615	695

Modeling Linear Data

Exercises 2-7: Refer to Examples 2-5 on pages 227-229 in your text and the Section 3.7 lecture video.

2.   The table shows the number of miles y traveled by a truck on x gallons of gasoline.

x	3	6	9	12
y	36	72	108	144

   (a) Plot the data in the xy-plane. Be sure to label each axis.

(b) Sketch a line that models the data.

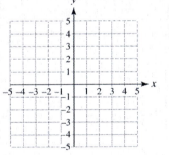

(c) Find the equation of the line and interpret the slope of the line.   2.(c)_____

(d) How far could this truck travel on 16 gallons of gasoline?   (d)_____

3. The table contains ordered pairs that can be modeled approximately by a line.

x	−3	−1	0	1	3
y	2	1	0	−1	−2

(a) Plot the data. Could a line pass through all five points?   3.(a)_____

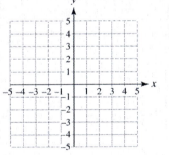

(b) Sketch a line that models the data and then determine its equation.   (b)_____

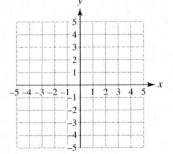

4. According to data from the 2011 National Diabetes Fact Sheet, at the beginning of 2011, a total of 18.8 million people had been diagnosed with diabetes, and the disease was growing at a rate of 1.9 million new diagnoses each year.

(a) Write a linear equation $N = mx + b$ that models the total number. **4.(a)**_____
of people N in millions that have been diagnosed with diabetes x years after January 1, 2011,

(b) Estimate N at the beginning of 2015. **(b)**_____

Find a linear equation in the form $y = mx + b$ that models the quantity y after x days.

5. A quantity y is initially 2500 and remains constant. **5.**_____

6. A quantity y is initially 700 and increases at a rate of 28 per day. **6.**_____

7. A quantity y is initially 480 and decreases at a rate of 10 per day. **7.**_____

Chapter 4 Systems of Linear Equations in Two Variables
4.1 Solving Systems of Linear Equations Graphically and Numerically

Basic Concepts ~ Solutions to Systems of Equations

STUDY PLAN

Read: Read Section 4.1 on pages 246-254 in your textbook or eText.

Practice: Do your assigned exercises in your ☐ Book ☐ MyMathLab ☐ Worksheets

Review: Keep your corrected assignments in an organized notebook and use them to review for the test.

Key Terms
Exercises 1-6: Use the vocabulary terms listed below to complete each statement. Note that some terms or expressions may not be used. Some terms may be used more than once.

parallel	inconsistent
identical	independent
consistent	solution to a system
dependent	intersection-of-graphs
intersecting	system of linear equations

1. A system of equations with exactly one solution is a(n) _____ system with _____ equations. Graphing the system results in _____lines.

2. The graphical technique of solving two equations is sometimes called the _____ method.

3. A system of equations with no solutions is a(n) _____ system. Graphing the system results in _____ lines.

4. A(n) _____ of two equations is an ordered pair (x, y) that makes both equations true.

5. A(n) _____ in two variables is written such that each equation is a linear equation in two variables.

6. A system of equations with infinitely many solutions is a(n) _____ system with _____ equations. Graphing the system results in _____ lines.

Basic Concepts

Exercises 1-3: Refer to Examples 1-2 on pages 247-248 in your text and the Section 4.1 lecture video.   ▪

1. The equation $P = 15x$ calculates an employee's pay for working   1. _____
    x hours at \$15 per hour. Use the intersection-of-graphs method
    to find the number of hours that the employee worked if the
    amount paid is \$75.

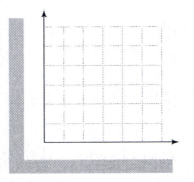

Use a graph to find the x-value when $y = -4$.

2. $y = -x - 2$ 2. _____

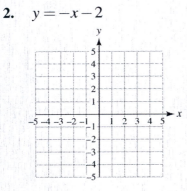

3. $4x - 3y = 12$ 3. _____

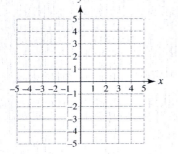

Solutions to Systems of Equations

Exercises 4-10: Refer to Examples 3-7 on pages 250-252 in your text and the Section 4.1 lecture video.

Graphs of two equations are shown. State the number of solutions to each system of equations. Then state whether the system is consistent or inconsistent. If it is consistent, state whether the equations are dependent or independent.

4.

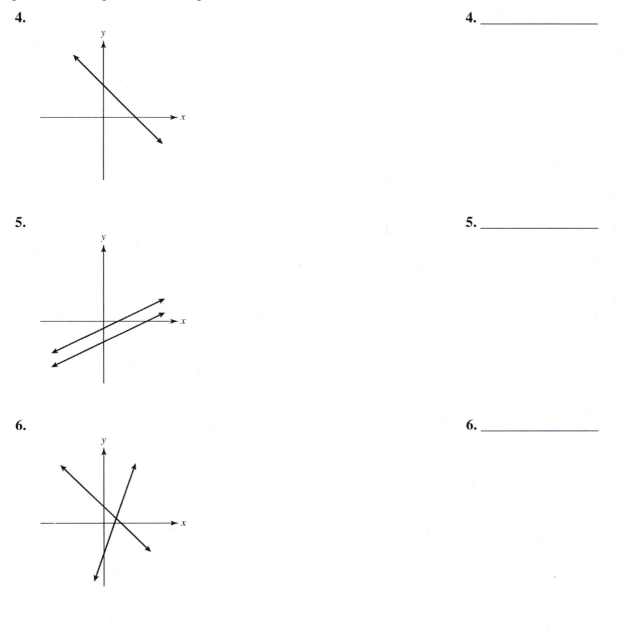

4. _____

5.

5. _____

6.

6. _____

7. Determine whether $(-5,-3)$ or $(-2,-2)$ is the solution to the   7. _____
    system of equations
    $$x - 2y = 1$$
    $$x - 3y = 4.$$

8. Solve the system of linear equations   8. _____
    $$x - y = -3$$
    $$3x - y = -3$$
    with a graph and with a table of values.

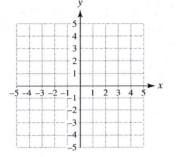

x				
$y =$				
$y =$				

9. Solve the system of equations graphically.   9. _____
    $$y = -2x$$
    $$3x - y = 5$$

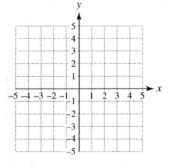

10. Last year, there were 12,000 students at a local college.   10. _____
    There were 2000 more female students than male students.
    How many male and female students attended the college?   _____

Chapter 4 Systems of Linear Equations in Two Variables
4.2 Solving Systems of Linear Equations by Substitution

The Method of Substitution ~ Recognizing Other Types of Systems ~ Applications

---

STUDY PLAN

   Read: Read Section 4.2 on pages 257-262 in your textbook or eText.

   Practice: Do your assigned exercises in your   ☐ Book ☐ MyMathLab ☐ Worksheets

   Review: Keep your corrected assignments in an organized notebook and use them to review for the test.

Key Terms
Exercises 1-3: Use the vocabulary terms listed below to complete each statement.
Note that some terms or expressions may not be used.

   parallel
   identical
   no solutions
   method of substitution
   infinitely many solutions

1. The technique of substituting an expression for a variable and solving the resulting equation is called the _____.

2. If the process of solving a system of equations results in a statement that is always false, like $7 = 4$, we conclude that the system has _____. Graphing the system results in _____ lines.

3. If the process of solving a system of equations results in a statement that is always true, like $-3 = -3$, we conclude that the system has _____. Graphing the system results in _____ lines.

The Method of Substitution

Exercises 1-4: Refer to Examples 1-2 on pages 258-259 in your text and the Section 4.2 lecture video.

Solve each system of equations.

1. $4x - y = -4$

$y = 2x$

1. _____

2. $x - 2y = 1$

$x = 4y$

2. _____

3. $x + y = 5$

$x - 4y = 5$

3. _____

4. $4x - y = 7$

$2x + 5y = 9$

4. _____

Recognizing Other Types of Systems

Exercises 5-6: Refer to Example 3 on page 260 in your text and the Section 4.2 lecture video.

If possible, use substitution to solve the system of equations. Then use graphing to help explain the result.

5. $x - 3y = 1$

        $2x - 6y = 2$

5. _____

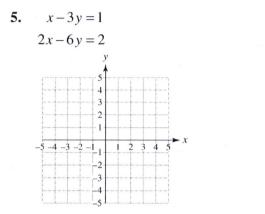

6. $2x - y = -2$

        $-6x + 3y = 4$

6. _____

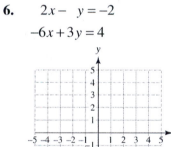

Applications

Exercises 7-9: Refer to Examples 4-6 on pages 260-262 in your text and the Section 4.2 lecture video.

7. A company places a total of 450 advertisements using radio and television. The number of radio ads is twice the number of television ads. Find the number of radio ads and the number of television ads.

7. _____

         _____

8. For a wedding reception, the hostess served desserts and hors d'oeuvres. A total of $2500 was spent on food. $700 more was spent on hors d'oeuvres than on desserts. How much was spent on hors d'oeuvres? How much was spent on desserts?

8. _____

_____

9. An airplane flies 1680 miles into (or against) the wind in 4 hours. The return trip takes $3\frac{1}{2}$ hours. Find the speed of the airplane with no wind and the speed of the wind.

9. _____

_____

Chapter 4 Systems of Linear Equations in Two Variables
4.3 Solving Systems of Linear Equations by Elimination

The Elimination Method ~ Recognizing Other Types of Systems ~ Applications

---

STUDY PLAN

Read: Read Section 4.3 on pages 265-273 in your textbook or eText.

Practice: Do your assigned exercises in your ☐ Book ☐ MyMathLab ☐ Worksheets

Review: Keep your corrected assignments in an organized notebook and use them to review for the test.

Key Terms
Exercises 1-3: Use the vocabulary terms listed below to complete each statement.
Note that some terms or expressions may not be used.

consistent	**independent**
no solutions	**inconsistent**
dependent	**infinitely many solutions**
elimination method	

1. If the process of solving a system of equations results in a statement that is always false, like $0 = 5$, we conclude that the system has _____ and is a(n) _____ system.

2. If the process of solving a system of equations results in a statement that is always true, like $0 = 0$, we conclude that the system has _____ and is a(n) _____ system with _____ equations.

3. The _____ is based on the addition property of equality.

The Elimination Method

Exercises 1-6: Refer to Examples 1-4 on pages 266-270 in your text and the Section 4.3 lecture video.

Solve each system of equations. Check each solution.

1. $x - 2y = -4$

    $x + 3y = 1$

1. _____

2. $3a - 2b = 2$

    $4a - 5b = -2$

2. _____

3. $-r - 4t = -8$

    $3r + \; t = -9$

3. _____

4. $2x + 6y = 4$

    $5x + 3y = 1$

4. _____

5. Solve the system of equations two times, first by eliminating x and then by eliminating y.

    $3y = -3 - 4x$
    $2x = -5y - 19$

5. _____

6. Solve the system of equations symbolically, graphically, and numerically.

    $x - y = 4$

    $4x + y = 1$

6. _____

x	-2	-1	0	1	2
$y =$					
$y =$					

Recognizing Other Types of Systems

Exercises 7-8: Refer to Example 5 on pages 270-271 in your text and the Section 4.3 lecture video.

Solve each system of equations by using the elimination method. Then graph the system.

7. $x - 4y = 6$ 7. _____

      $-x + 4y = -3$

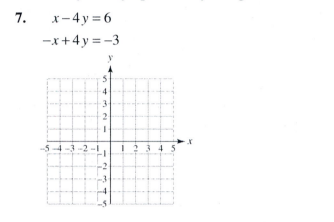

8. $x - 3y = 4$ 8. _____

      $-2x + 6y = -8$

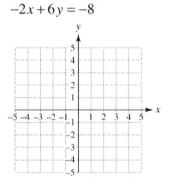

Applications

Exercises 9-10: Refer to Examples 6-7 on pages 271-272 in your text and the Section 4.3 lecture video.

9. Last year, enrollment at a local college increased by 1800
    students. There were 400 more new female students than
    new male students.

    (a) How many new students were women?   9. (a)_____

    (b) How many new students were men?   (b)_____

10. During strenuous exercise, an athlete can burn 12 calories   10. _____
    per minute on a treadmill and 8.5 calories per minute on
    an elliptical machine. If an athlete uses both machines and   _____
    burns 325 calories in a 30-minute workout, how many
    minutes does the athlete spend on each machine?

Understanding Concepts through Multiple Approaches
(For additional practice, visit MyMathLab.)

11. Solve the system of equations.

$$4x + 2y = 4$$
$$2x - y = -6$$

(a) Solve algebraically.

(b) Solve numerically using the table shown.

x	−2	−1	0	1	2
$y =$					
$y =$					

(c) Solve visually.

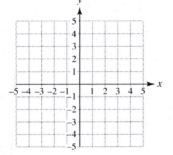

Did you get the same result using each method? Which method do you prefer? Explain why.

Chapter 4 Systems of Linear Equations in Two Variables
4.4 Systems of Linear Inequalities

Basic Concepts ~ Solutions to One Inequality ~ Solutions to Systems of Inequalities ~ Applications

---

STUDY PLAN

Read: Read Section 4.4 on pages 276-284 in your textbook or eText.

Practice: Do your assigned exercises in your ☐ Book ☐ MyMathLab ☐ Worksheets

Review: Keep your corrected assignments in an organized notebook and use them to review for the test.

Key Terms
Exercises 1-5: Use the vocabulary terms listed below to complete each statement.
Note that some terms or expressions may not be used.

> **solution**
> **test point**
> **solution set**
> **linear inequality**
> **system of linear inequalities**

1. A(n) _____ may be used to determine in which region of the xy-coordinate plane solutions to a linear inequality lie.

2. When the equals sign in any linear equation is replaced with $<$, $>$, $\leq$, or $\geq$, a(n) _____ in two variables results.

3. A(n) _____ to a linear inequality in two variables is an ordered pair (x, y) that makes the inequality a true statement.

4. A(n) _____ in two variables consists of two or more linear inequalities.

5. The _____ is the set of all solutions to an inequality.

Solutions to One Inequality

Exercises 1-5: Refer to Examples 1-2 on pages 278-279 in your text and the Section 4.4 lecture video.

Write a linear inequality that describes each shaded region.

1.

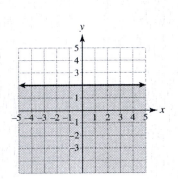

1. _____

2.

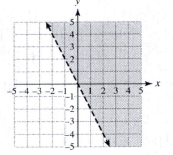

2. _____

Shade the solution set for each inequality.

3. $x \le 2$

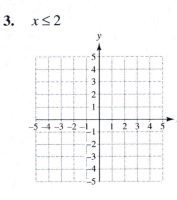

4. $y > -x + 5$

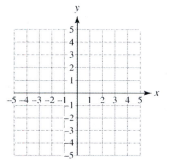

5. $2y - x \le 4$

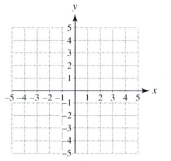

Solutions to Systems of Inequalities

Exercises 6-7: Refer to Examples 3-4 on pages 280-281 in your text and the Section 4.4 lecture video.

Shade the solution set to the system of inequalities.

6. $y \le x$
   $x + y > 2$

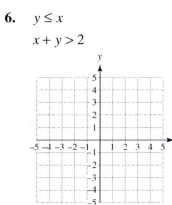

7. $x + 2y < -5$
    $2x - y \le 4$

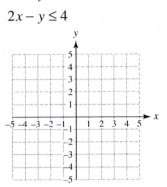

Applications

Exercises 8-9: Refer to Examples 5-6 on pages 282-283 in your text and the Section 4.4 lecture video.

8.   A business manufactures at least two MP3 players for every CD player. The total number of MP3 players and CD players is no more than 120. Shade the region that shows the number of MP3 players M and CD players P that can be produced within these restrictions. Label the horizontal axis P and the vertical axis M.

8. _____

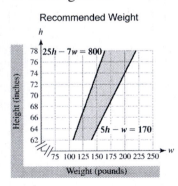

9.   The figure shows a shaded region containing recommended weights w for heights h.

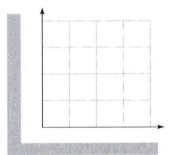

(a) What does the graph indicate about a 72-inch person who weighs 200 pounds?

9.(a)_____

(b) Determine the range of recommended weights for someone who is 74 inches tall.

(b)_____

Chapter 5 Polynomials and Exponents
5.1 Rules for Exponents

Review of Bases and Exponents ~ Zero Exponents ~ The Product Rule ~ Power Rules

---

STUDY PLAN

Read: Read Section 5.1 on pages 298-304 in your textbook or eText.

Practice: Do your assigned exercises in your ☐ Book ☐ MyMathLab ☐ Worksheets

Review: Keep your corrected assignments in an organized notebook and use them to review for the test.

Key Terms
Exercises 1-6: Use the vocabulary terms listed below to complete each statement.
Note that some terms or expressions may not be used.

exponent   undefined
power to a power   product
zero exponent   product to a power
base   quotient to a power

1. The _____ rule for exponents states that for any real number a and natural numbers m and n, $\left(a^m\right)^n = a^{m \cdot n}$.

2. The _____ rule for exponents states that for any real numbers a and b and natural number n, $\left(\dfrac{a}{b}\right)^n = \dfrac{a^n}{b^n}$, $b \neq 0$.

3. The _____ rule for exponents states that for any real number a and natural numbers m and n, $a^m \cdot a^n = a^{m+n}$.

4. The exponential expression 3^4 has _____ 3 and _____ 4.

5. The _____ rule for exponents states that for any real numbers a and b and natural number n, $\left(ab\right)^n = a^n b^n$.

6. The _____ rule states that for any nonzero real number b, $b^0 = 1$. The expression 0^0 is _____.

Review of Bases and Exponents

Exercises 1-4: Refer to Example 1 on pages 298-299 in your text and the Section 5.1 lecture video.

Evaluate each expression.

1. $3+\dfrac{2^3}{2}$

1. _____

2. $2\left(\dfrac{1}{2}\right)^3$

2. _____

3. -3^4

3. _____

4. $(-4)^2$

4. _____

Zero Exponents

Exercises 5-7: Refer to Example 2 on page 300 in your text and the Section 5.1 lecture video.

Evaluate each expression. Assume that all variables represent nonzero numbers.

5. -3^0

5. _____

6. $4\left(\dfrac{2}{3}\right)^0$

6. _____

7. $\left(\dfrac{xy^3}{2z}\right)^0$

7. _____

The Product Rule

Exercises 8-13: Refer to Examples 3-4 on pages 300-301 in your text and the Section 5.1 lecture video.

Multiply and simplify.

8. $3^2 \cdot 3^3$

8. _____

9. $x^6 x^2$

9. _____

10. $3x^4 \cdot 4x^3$

10. _____

11. $x^4 \left(3x^2 + 2x\right)$

11. _____

12. $a \cdot a^4$

12. _____

13. $(x+y)^3 (x+y)$

13. _____

Power Rules

Exercises 14-26: Refer to Examples 5-9 on pages 301-304 in your text and the Section 5.1 lecture video.

Simplify each expression.

14. $\left(2^3\right)^5$ 14. _____

15. $\left(x^4\right)^4$ 15. _____

16. $\left(5t\right)^3$ 16. _____

17. $\left(-2b^3\right)^3$ 17. _____

18. $3\left(x^3y^2\right)^4$ 18. _____

19. $\left(-2^2z^6\right)^3$ 19. _____

20. $\left(\dfrac{3}{4}\right)^3$ 20. _____

21. $\left(\dfrac{x}{y}\right)^6$

21. _____

22. $\left(\dfrac{6}{a-b}\right)^2$

22. _____

23. $(5x)^2(2x)^3$

23. _____

24. $\left(\dfrac{a^3b}{c^2}\right)^3$

24. _____

25. $(3xy^3)^2(-2x^4y^2)^3$

25. _____

26. If a parcel of property increases in value by about 16% each year for 15 years, then its value will triple two times.

 (a) Write an exponential expression that represents "tripling two times."

26. (a)_____

 (b) If the property is initially worth \$105,000, how much will it be worth if it triples 2 times?

 (b)_____

Chapter 5 Polynomials and Exponents
5.2 Addition and Subtraction of Polynomials

Monomials and Polynomials ~ Addition of Polynomials ~ Subtraction of Polynomials ~ Evaluating Polynomial Expressions

---

STUDY PLAN

    Read: Read Section 5.2 on pages 306-313 in your textbook or eText.

    Practice: Do your assigned exercises in your   ☐ Book   ☐ MyMathLab   ☐ Worksheets

    Review: Keep your corrected assignments in an organized notebook and use them to review for the test.

Key Terms
Exercises 1-10: Use the vocabulary terms listed below to complete each statement. Note that some terms or expressions may not be used. Some terms may be used more than once.

like	unlike
degree	monomial
binomial	coefficient
polynomial	trinomial

1. A polynomial with three terms is called a(n) _____.

2. The _____ of a monomial is the sum of the exponents of the variables.

3. A(n) _____ is a monomial or the sum of two or more monomials.

4. If two monomials contain the same variables raised to the same powers, we call them _____ terms.

5. A(n) _____ is a number, a variable, or a product of numbers and variables raised to natural number powers.

6. The expression $4x^3 - 3x^2 + 5x - 8$ is a(n) _____ in one variable.

7. The terms $7a^2b$ and $10ab^2$ are _____ terms.

8. The number in a monomial is called the _____ of the monomial.

9. A polynomial with two terms is called a(n) _____.

10. The _____ of a polynomial is the degree of the term (or monomial) with highest degree.

Monomials and Polynomials

Exercises 1-3: Refer to Example 1 on pages 307-308 in your text and the Section 5.2 lecture video.

Determine whether the expression is a polynomial. If it is, state how many terms and variables the polynomial contains and give its degree.

1. $2x^3 + 4x - 3$ 1. _____

2. $4x^3 - 2x^2 y^3 - xy^2 + 4y^3$ 2. _____

3. $3x^2 + \dfrac{1}{x-3}$ 3. _____

Addition of Polynomials

Exercises 4-10: Refer to Examples 2-5 on pages 309-310 in your text and the Section 5.2 lecture video.

State whether each pair of expressions contains like terms or unlike terms. If they are like terms, add them.

4. $5b^3; \ 2b^2$ 4. _____

5. $-3x^3 y; \ 7x^3 y$ 5. _____

6. $3xy^2; \ \dfrac{1}{2}xy^2$ 6. _____

Add by combining like terms.

7. $(2x+1)+(-3x+5)$ 7. _____

8. $(4y^2 - 3y + 7) + (y^2 + 6y + 1)$ 8. _____

9. Add $\left(7x^3 - 6x^2 + 2x - 4\right) + \left(5x^3 - 4x^2 + 3\right)$ by combining like terms.

                                                  9. _____

10. Simplify $\left(6b^2 + b + 7\right) + \left(-2b^2 - 5b + 6\right)$.

                                                  10. _____

Subtraction of Polynomials

Exercises 11-14: Refer to Examples 6-7 on page 311 in your text and the Section 5.2 lecture video.

Simplify each expression.

11. $\left(x + 5\right) - \left(-2x + 7\right)$

                                                  11. _____

12. $\left(4x^2 - 3x + 2\right) - \left(5x^2 + 7x + 1\right)$

                                                  12. _____

13. $\left(x^3 + 5x^2 - 8\right) - \left(-3x^3 + 6x - 4\right)$

                                                  13. _____

14. Simplify $\left(8x^2 - 3x - 1\right) - \left(5x^2 - 4x + 5\right)$.

                                                  14. _____

Evaluating Polynomial Expressions

Exercises 15-16: Refer to Examples 8-9 on pages 312-313 in your text and the Section 5.2 lecture video.

15. Write the monomial that represents the volume of a box having length x inches, width y inches, and height 4 inches. Find the volume of the box if $x = 8$ in. and $y = 10$ in.

15. _____

_____

16. The polynomial $0.835x - 1635$ approximates the price of a first-class postage stamp (in cents), where $x = 1980$ corresponds to the year 1980, $x = 1981$ to the year 1981, and so on. Estimate the price of a first-class postage stamp in the year 2002. Round answer to the nearest cent.

16. _____

Chapter 5 Polynomials and Exponents
5.3 Multiplication of Polynomials

Multiplying Monomials ~ Review of the Distributive Properties ~ Multiplying Monomials and Polynomials ~ Multiplying Polynomials

---

STUDY PLAN

Read: Read Section 5.3 on pages 316-322 in your textbook or eText.

Practice: Do your assigned exercises in your ☐ Book ☐ MyMathLab ☐ Worksheets

Review: Keep your corrected assignments in an organized notebook and use them to review for the test.

Key Terms
Exercise 1-3: Use the vocabulary terms listed below to complete each statement. Note that some terms or expressions may not be used. Some terms may be used more than once.

binomial	polynomial
variable	like term
product	coefficient

1. To multiply monomials in one variable, do the following.
    Step 1: Multiply the _____(s).
    Step 2: Use the product rule for exponents to multiply the _____(s).
    Step 3: Write the above results as a(n) _____.

2. The _____ of two polynomials may be found by multiplying every term in the first polynomial by every term in the second polynomial and then combining _____(s).

3. The FOIL process may be helpful for remembering how to multiply two _____(s).

Multiplying Monomials

Exercises 1-2: Refer to Example 1 on page 317 in your text and the Section 5.3 lecture video.

Multiply.

1. $-4x^3 \cdot 3x^2$

1. _____

2. $(4a^3b^2)(-5ab^4)$

2. _____

Review of the Distributive Properties

Exercises 3-5: Refer to Example 2 on page 317 in your text and the Section 5.3 lecture video.

Multiply.

3. $3(2x-1)$

3. _____

4. $(4x^2-5)2$

4. _____

5. $-x(5x-2)$

5. _____

Multiplying Monomials and Polynomials

Exercises 6-11: Refer to Examples 3-4 on page 318 in your text and the Section 5.3 lecture video.

Multiply.

6. $5x(3x^2-4)$

6. _____

7. $(4x-1)x^2$

7. _____

8. $-2\left(3x^2 - 4x + 2\right)$ **8.** _____

9. $3x^3\left(x^3 - 2x^2 + 3\right)$ **9.** _____

10. $4xy\left(5x^3y^2 + 1\right)$ **10.** _____

11. $-xy\left(x^2 + y^2\right)$ **11.** _____

Multiplying Polynomials

Exercises 12-20: Refer to Examples 5-9 on pages 319-321 in your text and the Section 5.3 lecture video.

12.   Multiply $(x+3)(x+5)$ geometrically and symbolically.   **12.** _____

Multiply. Draw arrows to show how each term is found.

13. $(2x-3)(x-1)$ **13.** _____

14. $(3-x)(1+4x)$ **14.** _____

15. $(3x-2)\left(x^2 + x\right)$ **15.** _____

Multiply.

16. $(4x+3)(x^2-x+1)$

16. _____

17. $(a+b)(4a^2-5b^2)$

17. _____

18. $(b^4-3b^2+1)(b^2-1)$

18. _____

19. Multiply x^2+3x+4 and $x-2$ vertically.

19. _____

20. A box has a width 2 inches less than its height and a length 5 inches more than its height.

(a) If h represents the height of the box, write a polynomial that represents the volume of the box.

20.(a)_____

(b) Use this polynomial to calculate the volume of the box if $h=15$ inches.

(b)_____

Chapter 5 Polynomials and Exponents
5.4 Special Products

Product of a Sum and Difference ~ Squaring Binomials ~ Cubing Binomials

---

STUDY PLAN

Read: Read Section 5.4 on pages 324-329 in your textbook or eText.

Practice: Do your assigned exercises in your   ☐ Book   ☐ MyMathLab   ☐ Worksheets

Review: Keep your corrected assignments in an organized notebook and use them to review for the test.

Key Terms
Exercise 1-3: Use the expressions listed below to complete each statement.
Note that some expressions may not be used.

$$a^2 + b^2$$
$$a^2 - b^2$$
$$a^2 + ab + b^2$$
$$a^2 + 2ab + b^2$$
$$a^2 - ab + b^2$$
$$a^2 - 2ab + b^2$$

1. For any real numbers a and b, $(a+b)(a-b) = $ _____.

2. For any real numbers a and b, $(a+b)^2 = $ _____.

3. For any real numbers a and b, $(a-b)^2 = $ _____.

Product of a Sum and Difference

Exercises 1-5: Refer to Examples 1-2 on pages 325-326 in your text and the Section 5.4 lecture video.

Multiply.

1. $(a-1)(a+1)$ 1. _____

2. $(z+3)(z-3)$ 2. _____

3. $(3r+5t)(3r-5t)$ 3. _____

4. $(3m^2+2n^2)(3m^2-2n^2)$ 4. _____

5. Use the product of a sum and difference to find $23 \cdot 17$.   5. _____

Squaring Binomials

Exercises 6-10: Refer to Examples 3-4 on pages 326-327 in your text and the Section 5.4 lecture video.

Multiply.

6. $(x-4)^2$ 6. _____

7. $(3x-2)^2$ 7. _____

8. $(1+6b)^2$ 8. _____

9. $(3a^2-b)^2$ 9. _____

10. A square pool has a 6-foot-wide walk around it.

    (a) If the sides of the pool have length x, find a polynomial   **10. (a)**_____
that gives the total area of the pool and sidewalk.

    (b) Let $x=20$ ft and evaluate the polynomial.   **(c)**_____

Cubing Binomials

Exercises 11-12: Refer to Examples 5-6 on pages 328-329 in your text and the Section 5.4 lecture video.

11. Multiply $(5x+2)^3$.   11. _____

12. If a savings account pays r percent annual interest, where r is expressed as a decimal, then after 3 years a sum of money will grow by a factor of $(1+r)^3$.

    (a) Multiply this expression.   **12. (a)**_____

    (b) Evaluate the expression for $r=0.04$ (or 4%), and interpret   **(b)**_____
the result. Round answer to the nearest thousandth.

Chapter 5 Polynomials and Exponents
5.5 Integer Exponents and the Quotient Rule

Negative Integers as Exponents ~ The Quotient Rule ~ Other Rules for Exponents ~ Scientific Notation

---

STUDY PLAN

   Read: Read Section 5.5 on pages 332-340 in your textbook or eText.

   Practice: Do your assigned exercises in your   ☐ Book   ☐ MyMathLab   ☐ Worksheets

   Review: Keep your corrected assignments in an organized notebook and use them to review for the test.

Key Terms
Exercise 1-4: Use the vocabulary terms listed below to complete each statement.
Note that some terms or expressions may not be used.

$$a^m - a^n \qquad \left(\frac{b}{a}\right)^n \qquad a^{m-n} \qquad \frac{b^m}{a^n}$$

$$\frac{1}{a^n} \qquad \frac{a^n}{b^m} \qquad \left(\frac{a}{b}\right)^n \qquad a^n$$

reciprocal   scientific notation

1. Let a be a nonzero real number and n be a positive integer. Then $a^{-n} =$ _____.
    That is, a^{-n} is the _____ of a^n.

2. For any nonzero number a and integers m and n, $\dfrac{a^m}{a^n} =$ _____.

3. The following three rules hold for any nonzero numbers a and b and positive integers m and n.

   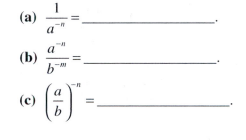

    (a) $\dfrac{1}{a^{-n}} =$ _____.

    (b) $\dfrac{a^{-n}}{b^{-m}} =$ _____.

    (c) $\left(\dfrac{a}{b}\right)^{-n} =$ _____.

4. A real number a is in _____ when a is written in the form $b \times 10^n$, where
    $1 \le |b| < 10$ and n is an integer.

Negative Integers as Exponents

Exercises 1-10: Refer to Examples 1-3 on pages 333-334 in your text and the Section 5.5 lecture video.

Simplify each expression.

1. 4^{-2} 1. _____

2. 5^{-1} 2. _____

3. x^{-3} 3. _____

4. $(a+b)^{-4}$ 4. _____

Evaluate each expression.

5. $3^2 \cdot 3^{-5}$ 5. _____

6. $5^{-3} \cdot 5^{-1}$ 6. _____

Simplify the expression. Write the answer using positive exponents.

7. $x^3 \cdot x^{-5}$ 7. _____

8. $\left(t^3\right)^{-2}$ 8. _____

9. $(ab)^{-4}$ 9. _____

10. $(xy)^{-5}\left(xy^{-3}\right)^2$ 10. _____

The Quotient Rule

Exercises 11-13: Refer to Example 4 on page 335 in your text and the Section 5.5 lecture video.

Simplify each expression. Write the answer using positive exponents.

11. $\dfrac{3^4}{3^6}$ 11. _____

12. $\dfrac{15a^2}{5a^6}$ 12. _____

13. $\dfrac{x^6 y^2}{x^5 y^7}$ 13. _____

Other Rules for Exponents

Exercises 14-17: Refer to Example 5 on page 336 in your text and the Section 5.5 lecture video.

Simplify each expression. Write the answer using positive exponents.

14. $\dfrac{1}{3^{-4}}$ 14. _____

15. $\dfrac{4^{-2}}{3^{-3}}$ 15. _____

16. $\dfrac{4x^2 y^{-3}}{12x^{-3} y^4}$ 16. _____

17. $\left(\dfrac{b^3}{5}\right)^{-2}$ 17. _____

Scientific Notation

Exercises 18-23: Refer to Examples 6-8 on pages 338-339 in your text and the Section 5.5 lecture video.

Write each number in standard form.

18. 4.17×10^5 18. _____

19. 2.3×10^{-4} 19. _____

20. 5.82×10^{-2} 20. _____

Write each number in scientific notation.

21. $32,000,000$ 21. _____

22. 0.00002 22. _____

23.   There are approximately 5.859×10^{12} miles in one light year.   23. _____
       The distance from Earth to the moon and back is 4.8×10^5 miles.
       How many round trips to the moon does one light-year represent?
       Write your answer in scientific notation.

Chapter 5 Polynomials and Exponents
5.6 Division of Polynomials

Division by a Monomial ~ Division by a Polynomial

Division by a Monomial

Exercises 1-5: Refer to Examples 1-3 on pages 343-345 in your text and the Section 5.6 lecture video.

Divide.

1. $\dfrac{x^6 - x^3}{x^2}$ 1. _____

2. $\dfrac{4b^8 - 12b^5}{8b^3}$ 2. _____

3. $\dfrac{15x^2 - 5x + 10}{5x}$ 3. _____

4. Divide the expression $\dfrac{6a^3 + 8a^2 - 4a}{2a^2}$ and then check the result.

4. _____

5. A rectangle has an area $A = x^2 + 3x$ and width x. Write an expression for its length l in terms of x.

5. _____

Division by a Polynomial

Exercises 6-8: Refer to Examples 4-6 on pages 345-347 in your text and the Section 5.6 lecture video.

6. Divide $\dfrac{6x^2 + 5x - 8}{2x + 3}$ and check.

6. _____

7. Simplify $\left(4x^3 - 2x + 5\right) \div \left(x - 1\right)$.

7. _____

8. Divide $x^3 - 4x^2 + 5x - 12$ by $x^2 + 4$.

8. _____

Chapter 6 Factoring Polynomials and Solving Equations
6.1 Introduction to Factoring

Basic Concepts ~ Common Factors ~ Factoring by Grouping

STUDY PLAN

Read: Read Section 6.1 on pages 360-366 in your textbook or eText.

Practice: Do your assigned exercises in your ☐ Book ☐ MyMathLab ☐ Worksheets

Review: Keep your corrected assignments in an organized notebook and use them to review for the test.

Key Terms
Exercises 1-3: Use the vocabulary terms listed below to complete each statement.
Note that some terms or expressions may not be used.

> **factor**
> **completely factored**
> **greatest common factor (GCF)**

1. We _____ a polynomial by writing the polynomial as a product of two or more lower degree polynomials.

2. The _____ of a list of positive integers is the largest integer that is a factor of every integer in the list.

3. A term is _____ when its coefficient is written as a product of prime numbers and any powers of variables are written as repeated multiplication.

Common Factors

Exercises 1-10: Refer to Examples 1-4 on pages 361-363 in your text and the Section 6.1 lecture video.

Factor the expression and sketch a rectangle that illustrates the factorization.

1. $30x + 18$

1. _____

2. $9x^2 - 12x$

2. _____

Factor.

3. $12x^2 + 8x$

3. _____

4. $10y^3 - 2y^2$

4. _____

5. $4a^3 - 12a^2 - 8a$

5. _____

6. $6x^3y - 2x^2y^2$

6. _____

Find the greatest common factor for each expression. Then factor the expression.

7. $4a^2 + 6a$

7. _____

_____

8. $5x^4 - 15x^2$

8. _____

_____

9. $6a^3b^2 - 15a^2b^3$

9. _____

_____

10. If a ball is thrown upward at 40 feet per second, then its
    height after t seconds is approximated by $40t - 16t^2$.
    Factor this expression.

10. _____

Factoring by Grouping

Exercises 11-18: Refer to Examples 5-8 on pages 363-365 in your text and the Section 6.1 lecture video.

Factor.

11. $4x(x-3) - 5(x-3)$

11. _____

12. $t^2(2t+3) + 6(2t+3)$

12. _____

Factor each polynomial.

13. $3x^3 - 6x^2 + 4x - 8$ 13. _____

14. $4x - 4y + ax - ay$ 14. _____

15. $5x^3 - 15x^2 - x + 3$ 15. _____

16. $2z^3 + 12z^2 - 3z - 18$ 16. _____

Completely factor each polynomial.

17. $8x^3 - 12x^2 + 8x - 12$ 17. _____

18. $4x^5 + 6x^4 - 10x^3 - 15x^2$ 18. _____

Chapter 6 Factoring Polynomials and Solving Equations
6.2 Factoring Trinomials I $(x^2 + bx + c)$

Review of the FOIL Method ~ Factoring Trinomials with Leading Coefficient 1

STUDY PLAN

Read: Read Section 6.2 on pages 368-373 in your textbook or eText.

Practice: Do your assigned exercises in your ☐ Book ☐ MyMathLab ☐ Worksheets

Review: Keep your corrected assignments in an organized notebook and use them to review for the test.

Key Terms
Exercises 1-2: Use the vocabulary terms listed below to complete each statement.
Note that some terms or expressions may not be used.

> **standard form**
> **prime polynomial**
> **leading coefficient**

1. Any trinomial of degree 2 in the variable x can be written in _____ as $ax^2 + bx + c$, where a, b, and c are constants. The constant a is called the

    _____ .

2. A polynomial with integer coefficients that cannot be factored by using integer coefficients is called a(n) _____ .

Factoring Trinomials with Leading Coefficient 1

Exercises 1-16: Refer to Examples 1-7 on pages 369-373 in your text and the Section 6.2 lecture video.

For each of the following, find an integer pair that has the given product and sum.

1.   Product: 28; Sum: 11   1. _____

2.   Product: -40; Sum: -3 2. _____

Factor each trinomial.

3. $x^2 + 7x + 10$ 3. _____

4. $x^2 + 9x + 18$ 4. _____

5. $y^2 + 13y + 42$ 5. _____

6. $b^2 - 10b + 21$ 6. _____

7. $x^2 - 8x + 12$ 7. _____

8. $y^2 - y - 20$ 8. _____

9. $t^2 - 3t - 40$ 9. _____

10. $x^2 + 2x - 24$ 10. _____

11. $x^2 - 7x + 12$ 11. _____

Factor each trinomial, if possible.

12. $x^2 - 9x + 22$ 12. _____

13. $x^2 - 5x - 14$ 13. _____

Factor each trinomial completely.

14. $5x^2 + 30x + 40$ 14. _____

15. $2x^4 + 10x^3 - 12x^2$ 15. _____

16.   Find one possibility for the dimensions of a rectangle   16. _____
that has an area of $x^2 + 3x - 10$.

Chapter 6   Factoring Polynomials and Solving Equations

6.3   Factoring Trinomials II $(ax^2 + bx + c)$

Factoring Trinomials by Grouping ~ Factoring with FOIL in Reverse

---

STUDY PLAN

   Read: Read Section 6.3 on pages 376-382 in your textbook or eText.

   Practice: Do your assigned exercises in your   ☐ Book   ☐ MyMathLab   ☐ Worksheets

   Review: Keep your corrected assignments in an organized notebook and use them to review
              for the test.

Factoring Trinomials by Grouping

Exercises 1-7:   Refer to Examples 1-3 on pages 377-379 in your text and the Section 6.3 lecture video.

Factor each trinomial.

   1. $2x^2 + 13x + 15$ **1.** _____

   2. $4a^2 - 11a - 3$ **2.** _____

   3. $10x^2 - 29x + 10$ **3.** _____

   4. $5x^2 + 9x - 5$ **4.** _____

   5. $2a^2 - 9a - 10$ **5.** _____

Factor each trinomial completely.

6. $15y^2 - 55y - 20$ 6. _____

7. $3x^3 - 18x^2 - 27x$ 7. _____

Factoring with FOIL in Reverse

Exercises 8-12: Refer to Examples 4-5 on pages 380-381 in your text and the Section 6.3 lecture video.

Factor each trinomial.

8. $3a^2 - 11a - 20$ 8. _____

9. $2x^2 + 11x + 5$ 9. _____

10. $8 + 5x^2 + 41x$ 10. _____

11. $-6x^2 - 13x + 5$ 11. _____

12. $3x + 35 - 2x^2$ 12. _____

Chapter 6 Factoring Polynomials and Solving Equations
6.4 Special Types of Factoring

Difference of Two Squares ~ Perfect Square Trinomials ~ Sum and Difference of Two Cubes

---

STUDY PLAN

Read: Read Section 6.4 on pages 384-389 in your textbook or eText.

Practice: Do your assigned exercises in your   ☐ Book ☐ MyMathLab ☐ Worksheets

Review: Keep your corrected assignments in an organized notebook and use them to review for the test.

Key Terms
Exercises 1-6: Use the expressions listed below to complete each statement.
Note that some expressions may not be used.

$a^2 + b^2$ $a^2 + 2ab + b^2$

$(a+b)^2$ $a^2 - 2ab + b^2$

$a^2 - b^2$ $(a-b)(a^2 + ab + b^2)$

$(a-b)^2$ $(a-b)(a^2 + 2ab + b^2)$

$(a+b)^3$ $(a+b)(a^2 - ab + b^2)$

$(a-b)^3$ $(a+b)(a^2 - 2ab + b^2)$

$(a-b)(a+b)$

1. For any real numbers a and b, $a^3 + b^3 = $ _____.

2. For any real numbers a and b, $a^2 - 2ab + b^2 = $ _____.

3. For any real numbers a and b, $a^3 - b^3 = $ _____.

4. For any real numbers a and b, $a^2 + 2ab + b^2 = $ _____.

5. For any real numbers a and b, $a^2 - b^2 = $ _____.

6. The expressions _____ and _____ are called perfect square trinomials.

Difference of Two Squares

Exercises 1-4: Refer to Example 1 on page 384 in your text and the Section 6.4 lecture video.

Factor each difference of two squares.

1. $x^2 - 16$

    1. _____

2. $25x^2 - 4$

    2. _____

3. $81 - 25a^2$

    3. _____

4. $9x^2 - 100y^2$

    4. _____

Perfect Square Trinomials

Exercises 5-8: Refer to Example 2 on pages 385-386 in your text and the Section 6.4 lecture video.

If possible, factor each trinomial as a perfect square trinomial.

5. $x^2 + 12x + 36$

    5. _____

6. $9t^2 + 6t + 1$

    6. _____

7. $25x^2 - 40x + 16$

    7. _____

8. $x^2 - 14xy + 49y^2$

    8. _____

Sum and Difference of Two Cubes

Exercises 9-16: Refer to Examples 3-5 on pages 387-388 in your text and the Section 6.4 lecture video.

Factor each polynomial.

9. $x^3 + 125$

9. _____

10. $a^3 - 8$

10. _____

11. $x^3 - 216$

11. _____

12. $64x^3 - 27$

12. _____

13. $16y^2 + 24y + 9$

13. _____

14. $16b^2 - 25$

14. _____

Factor each polynomial completely.

15. $12x^3 - 60x^2 + 75x$

15. _____

16. $9a^3 - 64ab^2$

16. _____

Chapter 6 Factoring Polynomials and Solving Equations
6.5 Summary of Factoring

Guidelines for Factoring Polynomials ~ Factoring Polynomials

STUDY PLAN

Read: Read Section 6.5 on pages 391-395 in your textbook or eText.

Practice: Do your assigned exercises in your   ☐ Book ☐ MyMathLab ☐ Worksheets

Review: Keep your corrected assignments in an organized notebook and use them to review for the test.

Key Terms
Exercises: Use the vocabulary terms listed below to complete each statement.
Note that some terms or expressions may not be used. Some terms may be used
more than once.

$a^2 + b^2$

$(a+b)^2$

$a^2 - b^2$

$(a-b)^2$

$a^3 + b^3$

$a^3 - b^3$

FOIL
grouping
perfect square
sum of two cubes
completely factored
difference of two squares
perfect square trinomial
greatest common factor (GCF)
difference of two cubes

Guidelines for Factoring Polynomials

STEP 1: Factor out the _____, if possible.

STEP 2: A. If the polynomial has four terms, try factoring by _____.

        B. If the polynomial is a binomial, try one of the following.

           1. _____ $=(a-b)(a+b)$ This is referred to as a(n) _____.

           2. _____ $=(a-b)(a^2+ab+b^2)$ This is referred to as a(n) _____.

           3. _____ $=(a+b)(a^2-ab+b^2)$ This is referred to as a(n) _____.

        C. If the polynomial is a trinomial, check for a(n) _____.

           1. $a^2 + 2ab + b^2 =$ _____ This is referred to as a(n) _____.

           2. $a^2 - 2ab + b^2 =$ _____ This is referred to as a(n) _____.

           Otherwise, try to factor the trinomial by _____ or apply _____ in reverse.

STEP 3: Check to make sure that the polynomial is _____.

Factoring Polynomials

Exercises 1-8: Refer to Examples 1-8 on pages 392-394 in your text and the Section 6.5 lecture video.

Factor.

1. $5x^3 - 20x^2 + 25x$ 1. _____

2. $4t^4 - 144t^2$ 2. _____

3. $-45a^3 - 30a^2 - 5a$ 3. _____

4. $5x^3 - 320$ 4. _____

5. $24x^4 + 10x^3 - 4x^2$ 5. _____

6. $8x^3 + 4x^2 - 72x - 36$ 6. _____

7. $16a^3b - 36ab^3$ 7. _____

8. $12x^3 + 9x^2 + 20x + 15$ 8. _____

Chapter 6 Factoring Polynomials and Solving Equations
6.6 Solving Equations by Factoring I (Quadratics)

The Zero-Product Property ~ Solving Quadratic Equations ~ Applications

---

STUDY PLAN

Read: Read Section 6.6 on pages 396-402 in your textbook or eText.

Practice: Do your assigned exercises in your ☐ Book ☐ MyMathLab ☐ Worksheets

Review: Keep your corrected assignments in an organized notebook and use them to review for the test.

Key Terms
Exercises 1-5: Use the vocabulary terms listed below to complete each statement.
Note that some terms or expressions may not be used.

> zeros
> standard form
> zero-product
> quadratic equation
> quadratic polynomial

1. The _____ property states that if the product of two numbers (or expressions) is 0, then at least one of the numbers (or expressions) must equal 0.

2. Any _____ in the variable x can be written as $ax^2 + bx + c$ with $a \neq 0$.

3. The _____ of a polynomial in x are the values that, when substituted for x, result in 0.

4. Any _____ in the variable x can be written as $ax^2 + bx + c = 0$ with $a \neq 0$.

5. The form $ax^2 + bx + c = 0$ is called the _____ of a quadratic equation.

The Zero-Product Property

Exercises 1-4: Refer to Example 1 on page 397 in your text and the Section 6.6 lecture video.

Solve each equation.

1. $x(x+2)=0$ 1. _____

2. $3a^2=0$ 2. _____

3. $(b+1)(b-4)=0$ 3. _____

4. $x(x-3)(x+5)=0$ 4. _____

Solving Quadratic Equations

Exercises 5-9: Refer to Examples 2-3 on pages 398-399 in your text and the Section 6.6 lecture video.

Solve each quadratic equation. Check your answers.

5. $x^2+4x=0$ 5. _____

6. $t^2=9$ 6. _____

7. $a^2-5a+6=0$ 7. _____

8. $10x^2 + 7x = 12$

8. _____

9. Solve $2x^2 - 9x = 5$.

9. _____

Applications

Exercises 10-12: Refer to Examples 4-6 on pages 400-401 in your text and the Section 6.6 lecture video.

10. The height h in feet of a baseball after t seconds is given by $h(t) = -16t^2 + 88t + 4$. At what values of t is the height of the baseball 100 feet?

10. _____

11. The braking distance D in feet required to stop a car traveling at x miles per hour on wet, level pavement can be approximated by $D = \dfrac{1}{9}x^2$.

    (a) Calculate the braking distance for a car traveling at 40 miles per hour.

11.(a)_____

    (b) If the braking distance is 60 feet, estimate the speed of the car.

(b)_____

    (c) If you have a calculator available, use it to solve part (b) numerically with a table of values.

(c)_____

12. A digital photograph is 20 pixels longer than it is wide and has a total area of 3500 pixels. Find the dimensions of this photograph.

12. _____

Understanding Concepts through Multiple Approaches
(For additional practice, visit MyMathLab.)

13. ***Braking Distance*** The braking distance D in feet required to stop a car traveling x miles per hour on wet, level pavement is approximated by $d = \frac{1}{9}x^2$. If the braking distance is 144 feet, estimate the speed of the car.

   (a)   Solve algebraically.

   (b)   Solve numerically using a calculator.

   Did you get the same result using each method? Which method do you prefer? Explain why.

Chapter 6 Factoring Polynomials and Solving Equations
6.7 Solving Equations by Factoring II (Higher Degree)

Polynomials with Common Factors ~ Special Types of Polynomials

---

STUDY PLAN

   Read: Read Section 6.6 on pages 405-408 in your textbook or eText.

   Practice: Do your assigned exercises in your ☐ Book ☐ MyMathLab ☐ Worksheets

   Review: Keep your corrected assignments in an organized notebook and use them to review for the test.

Polynomials with Common Factors

Exercises 1-5: Refer to Examples 1-3 on pages 405-407 in your text and the Section 6.7 lecture video.

Factor each trinomial completely.

   1. $-3x^2 + 9x + 12$ 1. _____

   2. $2x^3 - 12x^2 + 10x$ 2. _____

Solve each equation.

   3. $6y^3 - y^2 - y = 0$ 3. _____

   4. $4x^3 - 4x^2 = 120x$ 4. _____

5. A box is made by cutting a square with sides of length x from each corner of a rectangular piece of metal with length 30 inches and width 20 inches. The box has no top. If the outside surface area of the box is 500 square inches, find the value of x.

5. _____

Special Types of Polynomials

Exercises 6-10: Refer to Examples 4-5 on pages 407-408 in your text and the Section 6.7 lecture video.

Factor each polynomial completely.

6. $x^4 - 81$

6. _____

7. $a^4 + 6a^2 + 5$

7. _____

8. $r^4 - 2r^2t^2 + t^4$

8. _____

9. $a^4 - 16b^4$

9. _____

10. $x^4 - 64x$

10. _____

Name: _____   Course/Section: _____   Instructor: _____

Chapter 7 Rational Expressions
7.1 Introduction to Rational Expressions

Basic Concepts ~ Simplifying Rational Expressions ~ Applications

---

STUDY PLAN

Read: Read Section 7.1 on pages 420-427 in your textbook or eText.

Practice: Do your assigned exercises in your ☐ Book ☐ MyMathLab ☐ Worksheets

Review: Keep your corrected assignments in an organized notebook and use them to review for the test.

Key Terms
Exercises 1-6: Use the vocabulary terms listed below to complete each statement.
Note that some terms or expressions may not be used.

         undefined   **probability**
         lowest terms   **defined**
         basic principle   **rational expression**
         vertical asymptote

1. An expression is written in _____ when the numerator and denominator have no common factors.

2. The _____ of an event indicates the likelihood that the event will occur.

3. Division by 0 is _____.

4. A(n) _____ indicates a value of x at which a rational expression is undefined.

5. A(n) _____ can be written as $\dfrac{P}{Q}$, where P and Q are polynomials. It is _____ whenever $Q \neq 0$.

6. The _____ of rational expressions states that $\dfrac{P \cdot R}{Q \cdot R} = \dfrac{P}{Q}$ for $Q, R \neq 0$.

Basic Concepts

Exercises 1-8: Refer to Examples 1-2 on pages 420-421 in your text and the Section 7.1 lecture video.

If possible, evaluate each expression for the given value of the variable.

1. $\dfrac{1}{x-3}$ $x=2$

1. _____

2. $\dfrac{y^2}{3y-2}$ $y=-2$

2. _____

3. $\dfrac{3a+7}{a^2+2a+1}$ $a=-1$

3. _____

4. $\dfrac{x-4}{4-x}$ $x=1$

4. _____

Find all values of the variable for which each expression is undefined.

5. $\dfrac{1}{t}$

5. _____

6. $\dfrac{5x}{3x+2}$

6. _____

7. $\dfrac{1+4x}{x^2-9}$

7. _____

8. $\dfrac{3}{a^2+4}$

8. _____

Simplifying Rational Expressions

Exercises 9-17: Refer to Examples 3-5 on pages 421-424 in your text and the Section 7.1 lecture video.

Simplify each fraction by applying the basic principle of fractions.

9. $-\dfrac{8}{24}$

9. _____

10. $\dfrac{25}{40}$

10. _____

Simplify each expression.

11. $\dfrac{12t}{3t^2}$

11. _____

12. $\dfrac{3x-6}{4x-8}$

12. _____

13. $\dfrac{(x-2)(x+3)}{(x+3)(x-1)}$

13. _____

14. $\dfrac{x^2-25}{2x^2-11x+5}$

14. _____

15. $\dfrac{-t+4}{3t-12}$

15. _____

16. $\dfrac{6-x}{x-6}$

16._____

17. $-\dfrac{7-a}{a-7}$

17._____

Applications

Exercises 18-20: Refer to Examples 6-8 on pages 424-426 in your text and the Section 7.1 lecture video.

18. Suppose that 8 cars per minute can pass through a construction zone. If traffic arrives randomly at an average rate of x cars per minute, the average time T in minutes spent waiting in line and passing through the construction zone is given by $T = \dfrac{1}{8-x}$, where $x < 8$.

(a) Complete the table by finding T for each value of x.

x (cars/minute)	5	6	7	7.5	7.9	7.99
T (minutes)						

(b) Interpret the results.

18.(b)_____

19. Suppose that a small fish species is introduced into a pond that had not previously held this type of fish, and that its population P in thousands is modeled by $P = \dfrac{2x+1}{x+5}$, where $x \geq 0$ represents time in months.

(a) Complete the table by finding P for each value of x.
Round to 3 decimal places

x (months)	0	6	12	36	72
P (thousands)					

(b) How many fish were initially introduced into the pond?

19.(b)_____

(c) Interpret the results shown in your completed table.

(c)_____

20. Suppose that n balls, numbered 1 to n, are placed in a container and four balls have the winning number.

    (a) What is the probability of drawing the winning ball at random?

                                                      20.(a)_____

    (b) Calculate the probability for $n = 100, 1000$, and $10,000$.

                                                      (b)_____

    (c) What happens to the probability of drawing the winning ball as the number of balls increases?

                                                      (c)_____

Chapter 7 Rational Expressions
7.2 Multiplication and Division of Rational Expressions

Review of Multiplication and Division of Fractions ~ Multiplication of Rational Expressions ~ Division of Rational Expressions

---

STUDY PLAN

> **Read:** Read Section 7.2 on pages 431-434 in your textbook or eText.
>
> **Practice:** Do your assigned exercises in your ☐ Book ☐ MyMathLab ☐ Worksheets
> **Review:** Keep your corrected assignments in an organized notebook and use them to review for the test.

Key Terms
Exercises 1-4: Use the vocabulary terms listed below to complete each statement.
Note that some terms or expressions may not be used.

reciprocal	equivalent fraction
numerator	inverse principle of fractions
lowest term	basic principle of fractions
denominator	greatest common factor

1. To multiply two rational expressions, multiply the _____(s) and multiply the
   _____(s).

2. The _____ states that if the variables a, b, and c represent integers with $b \neq 0$ and
   $c \neq 0$, then $\dfrac{a \cdot c}{b \cdot c} = \dfrac{a}{b}$.

3. To divide two rational expressions, multiply by the _____ of the divisor.

4. A fraction is simplified to _____(s) if the GCF of its numerator and denominator
   is 1.

Review of Multiplication and Division of Fractions

Exercises 1-6: Refer to Examples 1-2 on page 431 in your text and the Section 7.2 lecture video.

Multiply and simplify your answers to lowest terms.

1. $\dfrac{2}{5} \cdot \dfrac{3}{7}$

1. _____

2. $3 \cdot \dfrac{5}{6}$

2. _____

3. $\dfrac{5}{18} \cdot \dfrac{3}{20}$

3. _____

Divide and simplify your answers to lowest terms.

4. $\dfrac{1}{5} \div \dfrac{3}{4}$

4. _____

5. $\dfrac{3}{7} \div 9$

5. _____

6. $\dfrac{7}{20} \div \dfrac{14}{15}$

6. _____

Multiplication of Rational Expressions

Exercises 7-11: Refer to Examples 3-4 on pages 432-433 in your text and the Section 7.2 lecture video.

Multiply and simplify to lowest terms. Leave your answers in factored form.

7. $\dfrac{4}{x} \cdot \dfrac{2x+3}{x+1}$

7._____

8. $\dfrac{x+2}{x-5} \cdot \dfrac{3x}{x+2}$

8._____

9. $\dfrac{x^2-9}{3x-1} \cdot \dfrac{3x-1}{x-3}$

9._____

10. $\dfrac{5}{x^2-3x+2} \cdot \dfrac{x^2+3x-4}{10}$

10._____

11. If a car is traveling at 60 miles per hour on a slippery road, then its stopping distance D in feet can be calculated by
$$D = \frac{3600}{30} \cdot \frac{1}{x},$$
where x is the coefficient of friction between the tires and the road and $0 < x \le 1$. The more slippery the road is, the smaller the value of x.

(a) Multiply and simplify the formula for D.

11.(a)_____

(b) Compare the stopping distance on an icy road with $x = 0.2$ to the stopping distance on dry pavement with $x = 0.6$.

(b)_____

Division of Rational Expressions

Exercises 12-14: Refer to Example 5 on pages 433-434 in your text and the Section 7.2 lecture video.

Divide and simplify to lowest terms.

12. $\dfrac{4}{3x} \div \dfrac{8}{x+2}$ 12. _____

13. $\dfrac{x^2-4}{x^2+1} \div (x+2)$ 13. _____

14. $\dfrac{x^2-3x}{x^2-2x-3} \div \dfrac{x}{x+3}$ 14. _____

Chapter 7 Rational Expressions
7.3 Addition and Subtraction with Like Denominators

Review of Addition and Subtraction of Fractions ~ Rational Expressions with Like Denominators

---

STUDY PLAN

Read: Read Section 7.3 on pages 437-442 in your textbook or eText.

Practice: Do your assigned exercises in your   ☐ Book   ☐ MyMathLab   ☐ Worksheets

Review: Keep your corrected assignments in an organized notebook and use them to review for the test.

Key Terms
Exercises 1-4: Use the vocabulary terms listed below to complete each statement. Note that some terms or expressions may not be used. Some terms may be used more than once.

greatest common factor	**numerator**
like denominator	**add**
equivalent fraction	**subtract**
denominator	

1. If two rational expressions have the same denominator, we say that they have
    _____(s).

2. To subtract two rational expressions having like denominators, _____ their
    _____(s). Keep the same _____.

3. The largest number that divides evenly into two or more given numbers is known as the
    _____.

4. To add two rational expressions having like denominators, _____ their
    _____(s). Keep the same _____.

Review of Addition and Subtraction of Fractions

Exercises 1-4: Refer to Example 1 on page 437 in your text and the Section 7.3 lecture video.

Simplify each expression to lowest terms.

1. $\dfrac{2}{7}+\dfrac{3}{7}$

1. _____

2. $\dfrac{3}{8}+\dfrac{3}{8}$

2. _____

3. $\dfrac{9}{4}-\dfrac{5}{4}$

3. _____

4. $\dfrac{17}{15}-\dfrac{7}{15}$

4. _____

Rational Expressions with Like Denominators

Exercises 5-17: Refer to Examples 2-6 on pages 438-441 in your text and the Section 7.3 lecture video.

Add and simplify to lowest terms.

5. $\dfrac{4}{t}+\dfrac{5}{t}$

5. _____

6. $\dfrac{x}{x+3}+\dfrac{3}{x+3}$

6. _____

7. $\dfrac{a-2}{a^2+5a}+\dfrac{2}{a^2+5a}$

7. _____

8. $\dfrac{x^2+x}{x+2}+\dfrac{3x+4}{x+2}$

8. _____

9. $\dfrac{3}{ab}+\dfrac{4}{ab}$

9. _____

10. $\dfrac{x}{x^2-y^2}+\dfrac{y}{x^2-y^2}$

10. _____

11. $\dfrac{2}{a-b}+\dfrac{-2}{b-a}$

11. _____

Subtract and simplify to lowest terms.

12. $\dfrac{x+3}{x}-\dfrac{3}{x}$

12. _____

13. $\dfrac{3y}{2y-5}-\dfrac{5y}{2y-5}$

13. _____

14. $\dfrac{x-4}{3x^2-x-4}-\dfrac{-5}{3x^2-x-4}$

14. _____

15. $\dfrac{4x}{3x+4}-\dfrac{x}{3x+4}$

15. _____

16. $\dfrac{a-b}{4b}-\dfrac{a+b}{4b}$

16. _____

17. A container holds a mixtures of TI-84 and TI-89 calculators. In this container, there is a total of n calculators, including 5 defective TI-84 calculators and 8 defective TI-89 calculators. If a calculator is picked at random by a quality control inspector, then the probability, or chance, of one of the defective calculators being chosen is given by the expression $\dfrac{5}{n}+\dfrac{8}{n}$.

   (a) Simplify this expression.

17.(a)_____

   (b) Interpret the result.

(b)_____

Chapter 7 Rational Expressions
7.4 Addition and Subtraction with Unlike Denominators

Finding Least Common Multiples ~ Review of Fractions with Unlike Denominators ~ Rational Expressions with Unlike Denominators

---

STUDY PLAN

   Read: Read Section 7.4 on pages 444-451 in your textbook or eText.

   Practice: Do your assigned exercises in your ☐ Book ☐ MyMathLab ☐ Worksheets

   Review: Keep your corrected assignments in an organized notebook and use them to review for the test.

Key Terms
Exercises 1-5: Use the vocabulary terms listed below to complete each statement.
Note that some terms or expressions may not be used.

   listing method   least common multiple
   greatest common factor   prime factorization method
   common multiple   least common denominator

1. Used to find the LCM of two or more expressions, the _____ involves finding the prime factorization of each expression. To find the LCM, list each factor the greatest number of times that it occurs in any one of the factorizations, and find the product of this list.

2. To add or subtract rational expressions with unlike denominators, we must first determine the _____ for all rational expressions involved.

3. A(n) _____ of two or more expressions is an expression that is divisible by each of the given expressions.

4. Used to find the LCM of two or more expressions, the _____ involves writing the multiples of each expression and choosing the least common multiple of these.

5. The _____ of two or more expressions is the simplest expression that is divisible by each of the expressions.

Finding Least Common Multiples

Exercises 1-5: Refer to Examples 1-2 on pages 444-445 in your text and the Section 7.4 lecture video.

Find the least common multiple of each pair of expressions.

1. $3a, 4a^2$ 1. _____

2. $z+1, z^2 + z$ 2. _____

3. $x-4, x+5$ 3. _____

4. $x^2 - 3x + 2, x^2 + 2x - 3$ 4. _____

5. Use a step diagram to find the LCM of $3x^3 - 15x^2$ and   5. _____
    $3x^3 - 12x^2 - 15x$.

Review of Fractions with Unlike Denominators

Exercises 6-7: Refer to Example 3 on page 446 in your text and the Section 7.4 lecture video.

Simplify each expression.

6. $\dfrac{5}{12} - \dfrac{7}{18}$ 6. _____

7. $\dfrac{2}{7} + \dfrac{3}{5}$ 7. _____

Rational Expressions with Unlike Denominators

Exercises 8-17: Refer to Examples 4-7 on pages 446-450 in your text and the Section 7.4 lecture video.

Rewrite each rational expression so it has the given denominator D.

8. $\dfrac{5}{3x}$, $D = 9x^2$

8. _____

9. $\dfrac{2}{x-y}$, $D = x^2 - y^2$

9. _____

Find each sum and leave your answer in factored form.

10. $\dfrac{4}{9y} + \dfrac{5}{18y^2}$

10. _____

11. $\dfrac{1}{x+2} + \dfrac{1}{x-2}$

11. _____

12. $\dfrac{x}{x^2+6x+9} + \dfrac{1}{x+3}$

12. _____

Simplify each expression. Write your answer in lowest terms and leave it in factored form.

13. $\dfrac{4}{a} - \dfrac{a}{a+2}$

13. _____

14. $\dfrac{3}{x-1} - \dfrac{2}{x^2-1}$

14. _____

15. $\dfrac{x}{x^2+x} - \dfrac{1}{x^2-x}$

15. _____

16. $\dfrac{4}{x} - \dfrac{1}{x+3} - \dfrac{3}{x^2+3x}$

16. _____

17. The length of a rectangle is $\dfrac{1}{x-5}$ and the width is $\dfrac{1}{x+1}$. Write an expression for the perimeter of the rectangle as a single rational expression in lowest terms.

17. _____

Chapter 7 Rational Expressions
7.5 Complex Fractions

Basic Concepts ~ Simplifying Complex Fractions

STUDY PLAN

Read: Read Section 7.5 on pages 454-460 in your textbook or eText.

Practice: Do your assigned exercises in your ☐ Book ☐ MyMathLab ☐ Worksheets

Review: Keep your corrected assignments in an organized notebook and use them to review for the test.

Key Terms
Exercises 1-4: Use the vocabulary terms listed below to complete each statement.
Note that some terms or expressions may not be used.

        reciprocal   **basic complex fraction**
        complex fraction   **greatest common factor**
        improper fraction   **least common denominator**

1. A(n) _____ is a rational expression that contains fractions in its numerator, denominator, or both.

2. Using Method II to simplify a complex fraction, we multiply both the numerator and the denominator of the complex fraction by the _____ of all fractions within the complex fraction.

3. Using Method I to simplify a complex fraction, we write the numerator and the denominator each as a single fraction and then multiply the fraction in the numerator by the _____ of the fraction in the denominator.

4. The expression $\dfrac{a}{b} \div \dfrac{c}{d}$ can be written as a(n) _____, where both the numerator and denominator are single fractions.

Simplifying Complex Fractions

Exercises 1-11: Refer to Examples 1-3 on pages 455-460 in your text and the Section 7.5 lecture video.

Simplify each basic complex fraction.

1. $\dfrac{\dfrac{2}{3}}{\dfrac{8}{9}}$

1. _____

2. $\dfrac{2\dfrac{5}{8}}{3\dfrac{3}{4}}$

2. _____

3. $\dfrac{\dfrac{x}{5}}{\dfrac{2}{5y}}$

3. _____

4. $\dfrac{\dfrac{(z+2)^2}{3}}{\dfrac{(z+2)}{9}}$

4. _____

Simplify. Write your answer in lowest terms.

5. $\dfrac{\dfrac{1}{x}-\dfrac{1}{y}}{\dfrac{1}{x}+\dfrac{1}{y}}$

5. _____

6. $\dfrac{a+\dfrac{3}{a}}{a-\dfrac{3}{a}}$

6. _____

7. $\dfrac{\dfrac{3}{x-2}+\dfrac{1}{x}}{\dfrac{3}{x}-\dfrac{1}{x-2}}$

7. _____

8. $\dfrac{\dfrac{1}{a}+\dfrac{1}{b}}{\dfrac{1}{a^2}-\dfrac{1}{b^2}}$

8. _____

Simplify.

9. $\dfrac{\dfrac{5}{x} - \dfrac{3}{x}}{2x}$

9. _____

10. $\dfrac{\dfrac{1}{x+2}}{\dfrac{4}{x+2} + \dfrac{1}{x}}$

10. _____

11. $\dfrac{\dfrac{1}{x} - \dfrac{1}{y}}{\dfrac{1}{3x^2} - \dfrac{1}{3y^2}}$

11. _____

Chapter 7 Rational Expressions
7.6 Rational Equations and Formulas

Solving Rational Equations ~ Rational Expressions and Equations ~ Graphical and Numerical Solutions ~ Solving a Formula for a Variable ~ Applications

---

STUDY PLAN

Read: Read Section 7.6 on pages 463-472 in your textbook or eText.

Practice: Do your assigned exercises in your ☐ Book ☐ MyMathLab ☐ Worksheets

Review: Keep your corrected assignments in an organized notebook and use them to review for the test.

Key Terms
Exercises 1-6: Use the vocabulary terms listed below to complete each statement. Note that some terms or expressions may not be used. Some terms may be used more than once.

term
visually
extraneous solution
rational equation
algebraically

greatest common factor
basic rational equation
numerically
least common denominator

1. If an equation contains one or more rational expressions, it is called a(n) _____.

2. When we solve an equation using the distributive property and the properties of equality, we are solving the equation _____.

3. To solve a rational equation, do the following.
    Step 1: Find the _____ of the terms in the equation.
    Step 2: Multiply each side of the equation by the _____.
    Step 3: Simplify each _____.
    Step 4: Solve the resulting equation.
    Step 5: Check each answer in the given equation. Any value that makes a denominator equal 0 should be rejected because it is a(n) _____.

4. When we solve an equation by estimating values from its graph, we are solving the equation _____.

5. A(n) _____ has a single rational expression on each side of the equals sign.

6. When we solve an equation using a table of values, we are solving the equation _____.

Solving Rational Equations

Exercises 1-7: Refer to Examples 1-3 on pages 464-466 in your text and the Section 7.6 lecture video.

Solve each equation.

1. $\dfrac{4}{3} = \dfrac{5}{x}$

1. _____

2. $\dfrac{x-2}{3} = \dfrac{4x}{3}$

2. _____

3. $\dfrac{4}{2x-7} = x$

3. _____

4. $\dfrac{3}{x} - \dfrac{5}{7} = \dfrac{2}{x}$

4. _____

Solve each equation. Check your answer.

5. $\dfrac{1}{a+2} - \dfrac{1}{a} = \dfrac{1}{4a}$

5. _____

6. $\dfrac{1}{x+2}+\dfrac{1}{x-2}=\dfrac{5}{x^2-4}$

6. _____

7. If possible, solve $\dfrac{2}{x+3}+\dfrac{1}{x-3}=\dfrac{6}{x^2-9}$.

7. _____

Rational Expressions and Equations

Exercises 8-9: Refer to Example 4 on pages 467-468 in your text and the Section 7.6 lecture video.

Determine whether you are given an expression or an equation. If it is an expression, simplify it and then evaluate it for $x = 4$. If it is an equation, solve it.

8. $\dfrac{x^2-3}{x+2}+\dfrac{1}{x+2}$

8. _____

 _____

9. $\dfrac{x+1}{x-3}=\dfrac{x}{x-1}$

9. _____

 _____

Graphical and Numerical Solutions

Exercise 10: Refer to Example 5 on pages 468-469 in your text and the Section 7.6 lecture video.

10.   Solve $\dfrac{3}{x} = x - 2$ graphically and numerically.

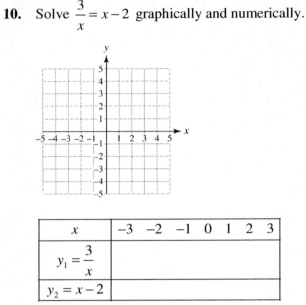

x	-3	-2	-1	0	1	2	3
$y_1 = \dfrac{3}{x}$							
$y_2 = x - 2$							

Solving a Formula for a Variable

Exercises 11-14: Refer to Examples 6-7 on pages 469-470 in your text and the Section 7.6 lecture video.

11.   If a person travels at a speed, or rate, r for time t, then the distance traveled is $d = rt$.

   (a) How far does a person travel in 2.5 hours when traveling   11. (a)_____
   at 60 miles per hour?

   (b) Solve the formula $d = rt$ for t.   (b)_____

   (c) How long does it take a person to go 250 miles when traveling   (c)_____
   at 75 miles per hour?

Solve each equation for the specified variable.

12. $A = \dfrac{bh}{2}$ for h

12. _____

13. $P = \dfrac{nRT}{V}$ for T

13. _____

14. $h = \dfrac{2A}{b_1 + b_2}$ for b_2

14. _____

Applications

Exercises 15-16: Refer to Examples 8-9 on pages 470-472 in your text and the Section 7.6 lecture video.

15.   One person can clean a carpet in 2 hours. A second person requires 3 hours to clean the carpet. How long will it take the two people, working together, to clean the carpet?

15. _____

16.   A runner is competing in a 10-mile race. He runs the first 8 miles at 6 miles per hour, and the last two miles at 7.5 miles per hour. How long does it take him to finish the race?

16. _____

Understanding Concepts through Multiple Approaches
(For additional practice, visit MyMathLab.)

17. Solve the equation $\dfrac{2}{x-3} = 2$.

(a) Solve algebraically.

(b) Solve numerically using the table shown.

x	-2	0	2	4	6
$y = \dfrac{2}{x-3}$					
$y = 2$					

(c) Solve visually.

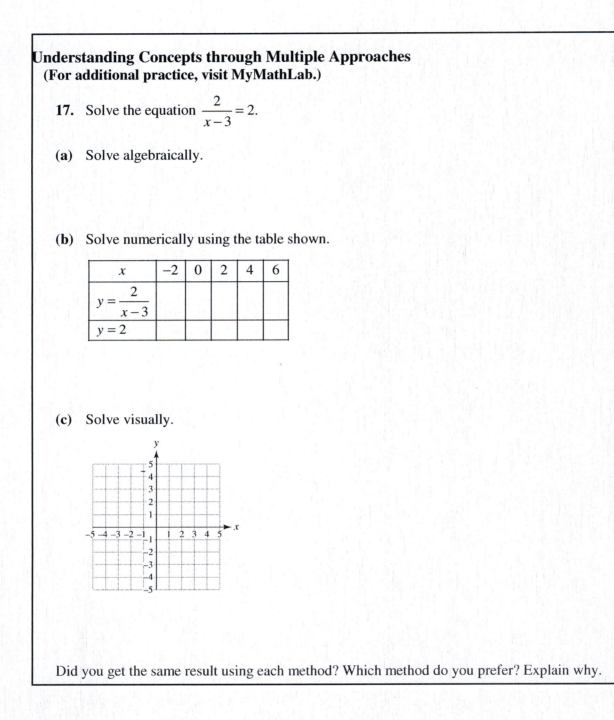

Did you get the same result using each method? Which method do you prefer? Explain why.

Chapter 7 Rational Expressions
7.7 Proportions and Variation

Proportions ~ Direct Variation ~ Inverse Variation ~ Analyzing Data ~ Joint Variation

---

STUDY PLAN

Read: Read Section 7.7 on pages 476-486 in your textbook or eText.

Practice: Do your assigned exercises in your   ☐ Book ☐ MyMathLab ☐ Worksheets

Review: Keep your corrected assignments in an organized notebook and use them to review for the test.

Key Terms
Exercises 1-6: Use the vocabulary terms listed below to complete each statement.
Note that some terms or expressions may not be used.

ratio   directly proportional
varies inversely   varies jointly
proportion   constant of proportionality
varies directly   inversely proportional
constant of variation

1. Let x and y denote two quantities. Then y is _____ to x, or y

    _____ with x, if there is a nonzero number k such that $y = \dfrac{k}{x}$.

2. A(n) _____ is a comparison of two quantities.

3. Let x and y denote two quantities. Then y is _____ to x, or y
    _____ with x, if there is a nonzero number k such that $y = kx$.

4. A(n) _____ is a statement that two ratios are equal.

5. Let x, y, and z denote three quantities. Then z _____ with x and y
    if there is a nonzero number k such that $z = kxy$.

6. For direct, inverse, or joint variation formulas, the number k is called the _____,
    or the _____.

Proportions

Exercises 1-2: Refer to Examples 1-2 on pages 477-478 in your text and the Section 7.7 lecture video.

1. Eight inches of heavy, wet snow are equivalent to one inch of rain   1. _____
    in terms of water content. If 14 inches of this type of snow fall,
    estimate the water content.

2. A 6-foot-tall person casts a 9-foot-long shadow. If a nearby tree   2. _____
    casts a 24-foot-long shadow, estimate the height of the tree.

Direct Variation

Exercises 3-5: Refer to Examples 3-5 on pages 479-481 in your text and the Section 7.7 lecture video.

3. Let y be directly proportional to x, or vary directly with x.   3. _____
    Suppose $y = 5$ when $x = 6$. Find y when $x = 8$.

4. The cost of tuition is directly proportional to the number of credits   4. _____
    taken. If 8 credits cost $944, find the cost to take 14 credits.

5. (a) A scatterplot of data is shown in the following figure.   5.(a)_____
    Could the data be modeled using a line?

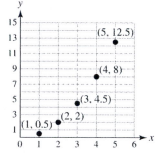

    (b) Is y directly proportional to x?   (b)_____

Inverse Variation

Exercises 6-7: Refer to Examples 6-7 on pages 482-484 in your text and the Section 7.7 lecture video.

6. Let y be inversely proportional to x, or vary inversely with x.   6. _____
    Suppose $y = 10$ when $x = 4$. Find y when $x = 8$.

7. (a) A scatterplot of data is shown in the following figure.   7.(a)_____
    Could the data be modeled using a line?

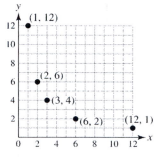

    (b) Is y inversely proportional to x?   (b)_____

    (c) Predict the y value if $x = 4$.   (c)_____

Analyzing Data

Exercises 8-10: Refer to Example 8 on page 484 in your text and the Section 7.7 lecture video.

Determine whether the data in each table represent direct variation, inverse variation, or neither.

8.

x	3	4	8	12
y	8	6	3	2

8. _____

9.

x	25	16	9	4
y	5	4	3	2

9. _____

10.

x	1	3	5	6
y	3	9	15	18

10. _____

Joint Variation

Exercise 11: Refer to Example 9 on page 485 in your text and the Section 7.7 lecture video.

11.   The strength S of a rectangular beam varies jointly with its width w and the square of its thickness t. If a beam 5 inches wide and 4 inches thick supports 400 pounds, how much can a similar beam 6 inches wide and 2 inches thick support?

11. _____

Chapter 8 Introduction to Functions
8.1 Functions and Their Representations

Basic Concepts ~ Representations of a Function ~ Definition of a Function ~ Identifying a Function ~ Graphing Calculators (Optional)

STUDY PLAN

Read: Read Section 8.1 on pages 503-516 in your textbook or eText.

Practice: Do your assigned exercises in your ☐ Book ☐ MyMathLab ☐ Worksheets

Review: Keep your corrected assignments in an organized notebook and use them to review for the test.

Key Terms
Exercises 1-6: Use the vocabulary terms listed below to complete each statement.
Note that some terms or expressions may not be used.

input	domain
range	output
function	relation
vertical line test	nonlinear
function notation	independent
dependent	name of the function

1. A(n) _____ is a set of ordered pairs.

2. A(n) _____ f is a set of ordered pairs (x, y), where each x-value corresponds to exactly one y-value. The _____ of f is the set of all x-values, and the _____ of f is the set of all y-values.

3. The notation $y = f(x)$ is called _____. The _____ is x, the _____ is y, and the _____ is f.

4. A function whose graph is not a line is called a(n) _____ function.

5. Using the notation $y = f(x)$, the variable y is called the _____ variable and the variable x is called the _____ variable.

6. To determine whether or not a graph represents a function, we can use the _____.

Representations of a Function

Exercises 1-6: Refer to Examples 1-4 on pages 506-508 in your text and the Section 8.1 lecture video.

Evaluate each function f at the given value of x.

1. $f(x) = 2x - 9, \ x = -3$ 1. _____

2. $f(x) = \dfrac{-3x}{x-1}, \ x = 4$ 2. _____

3. $f(x) = \sqrt{6x-2}, \ x = 3$ 3. _____

4. Let a function f compute a sales tax of 5% on a purchase of x dollars. Use the given representation to evaluate $f(3)$.

   (a) Verbal Representation: Multiply a purchase of x dollars by   4.(a)_____
   0.07 to obtain a sales tax of y dollars.

   (b) Numerical Representation   (b)_____

x	$f(x)$
$1.00	$0.05
$2.00	$0.10
$3.00	$0.15
$4.00	$0.20

   (c) Symbolic Representation $f(x) = 0.05x$ (c)_____

   (d) Graphical Representation   (d)_____

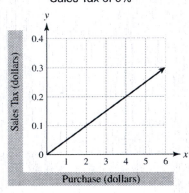

Sales Tax of 5%

4. **(e)** Diagrammatic Representation

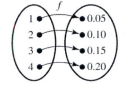

4.(d)_____

5. The number of World Wide Web searches S in billions during year x can be approximated by $S(x) = 225x - 450,650$ from 2009 to 2012.

(a) Find $S(2010)$ and interpret the result.

5.(a)_____

(b) How many more searches will there be in 2012 than in 2011?

(b)_____

6. Let function f square the input x and then subtract 4 to obtain the output y.

(a) Write a formula, or symbolic representation, for f.

6.(a)_____

(b) Make a table of values, or numerical representation, for f. Use $x = -2, -1, 0, 1, 2$.

x	-2	-1	0	1	2
$f(x)$					

(c) Sketch a graph, or graphical representation, of f.

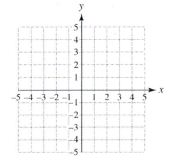

Definition of a Function

Exercises 7-12: Refer to Examples 5-7 on pages 509-511 in your text and the Section 8.1 lecture video.

7. The function f computes the average score on an algebra exam by the number of hours of study time. This function is defined by $f(1) = 68$, $f(3) = 77$, $f(5) = 84$, and $f(7) = 92$.

    (a) Write f as a set of ordered pairs.   7.(a)_____

    (b) Give the domain and range of f.   (b)_____

    (c) Discuss the relationship between study time and exam score.   (c)_____

Use the graphs of f shown to find each function's domain and range.

8.   8. _____

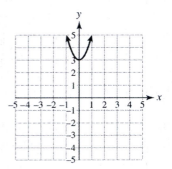

9.   9. _____

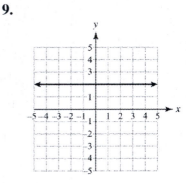

Use f(x) to find the domain of f.

10. $f(x) = 2x + 5$ **10.** _____

11. $f(x) = \dfrac{x-3}{x}$ **11.** _____

12. $f(x) = \sqrt{x-4}$ **12.** _____

Identifying a Function

Exercises 13-17: Refer to Examples 8-10 on pages 511-513 in your text and the Section 8.1 lecture video.

13.   The set S of ordered pairs (x, y) represents the zip code y for   **13.** _____
selected students x.

$S = \{(\text{Sam}, 84091), (\text{Erin}, 83705), (\text{Matt}, 59106), (\text{Niles}, 55421), (\text{Katie}, 84091), (\text{Becca}, 47552)\}$
Determine if S is a function.

14.   Determine whether the data in the table represents a function.   **14.** _____

x	2	1	−1	4	−1
y	0	−8	−2	−7	3

Determine whether the graphs shown represent functions.

15.

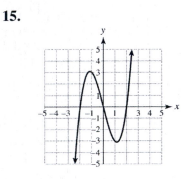

15. _____

16.

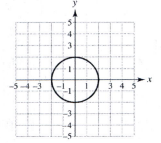

16. _____

17.

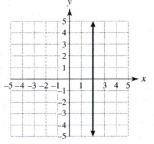

17. _____

Graphing Calculators (Optional)

Exercises 18-19: Refer to Examples 11-12 on pages 514-515 in your text and the Section 8.1 lecture video.

18. Show the viewing rectangle $[-5, 10, 0.75]$ by $[-50, 50, 5]$ on your calculator.

18. _____

19. Plot the points $(-3, -3)$, $(-2, 1)$, $(0, -2)$, and $(2, 4)$ in $[-5, 5, 1]$ by $[-5, 5, 1]$.

19. _____

Chapter 8 Introduction to Functions
8.2 Linear Functions

Basic Concepts ~ Representations of Linear Functions ~ Modeling Data with Linear Functions ~ The Midpoint Formula (Optional)

STUDY PLAN

Read: Read Section 8.2 on pages 522-534 in your textbook or eText.

Practice: Do your assigned exercises in your ☐ Book ☐ MyMathLab ☐ Worksheets

Review: Keep your corrected assignments in an organized notebook and use them to review for the test.

Key Terms
Exercises 1-5: Use the vocabulary terms listed below to complete each statement. Note that some terms or expressions may not be used.

midpoint	y-intercept
linear function	constant function
rate of change	nonlinear function

1. A function f defined by $f(x) = mx + b$, where m and b are constants, is a(n)

    _____.

2. The value of m represents the _____ of the graph of $f(x) = mx + b$, and b is the _____.

3. $f(x) = x^2 + 4$ is an example of a(n) _____.

4. The _____ of a line segment is the unique point on the line segment that is an equal distance from the endpoints.

5. A(n) _____ is a linear function with $m = 0$ and can be written as $f(x) = b$.

Basic Concepts

Exercises 1-7: Refer to Examples 1-2 on page 525 in your text and the Section 8.2 lecture video.

Determine whether f is a linear function. If f is a linear function, find values for m and b so that $f(x) = mx + b.$

1. $f(x) = 3 - \dfrac{1}{2}x$ 1. _____

2. $f(x) = 4x$ 2. _____

3. $f(x) = \dfrac{3}{x} - 7$ 3. _____

Use each table of values to determine whether $f(x)$ *could represent a linear function. If f could be linear, write a formula for f in the form* $f(x) = mx + b.$

4.
x	−2	−1	0	1	2
$f(x)$	−5	−3	−1	1	3

4. _____

5.
x	−2	−1	0	1	2
$f(x)$	4	1	0	1	4

5. _____

6.
x	−1	0	1	2	3
$f(x)$	−2	−2	−2	−2	−2

6. _____

7.
x	−2	−1	0	1	2
$f(x)$	2	1	0	−1	−2

7. _____

Representations of Linear Functions

Exercises 8-10: Refer to Examples 3-5 on pages 526-527 in your text and the Section 8.2 lecture video.

8. Sketch a graph of $f(x) = -x - 2$. Use the graph to evaluate $f(-4)$.   8. _____

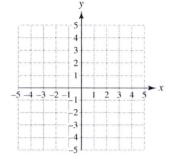

9. Give numerical and graphical representations of $f(x) = -\dfrac{1}{2}x + 1$.   9. _____

10. A linear function is given by $f(x) = 2x - 1$.

    (a) Give a verbal representation of f.   10.(a)_____

    (b) Make a numerical representation (table) of f by letting   (b)_____
    $x = -1, 0, 1$.

x	-1	0	1
$f(x)$			

    (c) Plot the points listed in the table from part (b).   (c)_____
    Then sketch a graph of $y = f(x)$.

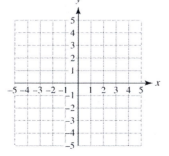

Modeling Data with Linear Functions

Exercises 11-13: Refer to Examples 6-8 on pages 528-530 in your text and the Section 8.2 lecture video.

11. The total cost to manufacture an item includes fixed costs of $1200 and per item cost of $75.

    (a) Find a function C that models total cost to manufacture x items.   11.(a)_____

    (b) Find $C(150)$ and interpret the result.   (b)_____

12. The table shows recommended cumulative weight loss over a 5-week period.

Week	0	1	2	3	4	5
Weight Loss (in lbs)	0	2	4	6	8	10

    (a) Make a scatterplot of the data and sketch the graph of a function f that models these data. Let x represent the number of weeks that have passed.

    (b) What is the recommended cumulative weight loss after 4 weeks? What is the recommended weekly weight loss?   12.(b)_____

    _____

    (c) Find a formula for $f(x)$.   (c)_____

    (d) Use your formula to estimate the recommended cumulative weight loss after 12 weeks.   (d)_____

13. An athlete runs and records his pace at regular intervals, as listed in the table.

Elapsed Time (minutes)	10	15	20	25	30
Pace (miles per hour)	8	8	8	8	8

(a) Discuss the pace of the athlete during this time interval.

13.(a)_____

(b) Find a formula for a function f that models these data.

(b)_____

(c) Sketch a graph of f together with the data.

(c)_____

The Midpoint Formula (Optional)

Exercises 14-16: Refer to Examples 9-11 on pages 532-533 in your text and the Section 8.2 lecture video.

14. Find the midpoint of the line segment connecting the points $(-1,-5)$ and $(2,1)$.

14. _____

15. Divorce rates in the United States vary state to state. In the year 2010, Massachusetts had the lowest rate at 1.8 per 1000 people, and Nevada had the highest rate at 6.6 per 1000 people.

(a) Find the midpoint of these data.

15.(a)_____

(b) Does the midpoint reflect the overall U.S. divorce rate of 3.4 per 1000 people?

(b)_____

16. The graph of a linear function f shown in the figure passes through the points $(-3,4)$ and $(2,-1)$.

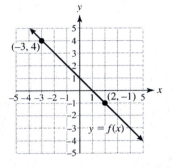

(a) Find a formula for $f(x)$.   **16.(a)**_____

(b) Evaluate $f(0)$. Does your answer agree with the graph?   **(b)**_____

(c) Find the midpoint M of the line segment connecting the points   **(c)**_____
$(-3,4)$ and $(2,-1)$. Comment on your result.

Chapter 8 Introduction to Functions
8.3 Compound Inequalities

Basic Concepts ~ Symbolic Solutions and Number Lines ~ Numerical and Graphical Solutions ~ Interval Notation

---

STUDY PLAN

Read: Read Section 8.3 on pages 540-547 in your textbook or eText.

Practice: Do your assigned exercises in your ☐ Book ☐ MyMathLab ☐ Worksheets

Review: Keep your corrected assignments in an organized notebook and use them to review for the test.

Key Terms
Exercises 1-6: Use the vocabulary terms listed below to complete each statement.
Note that some terms or expressions may not be used.

union	**set-builder**
interval	**intersection**
three-part	$A \cup B$
compound	$A \cap B$

1. A(n) _____ inequality consists of two inequalities joined by the word *and* or *or*.

2. For any two sets A and B, the _____ of A and B, denoted _____, is defined as $\{x \mid x$ is an element of A *or* an element of $B\}$.

3. The expression $\{x \mid -1 < x \le 2\}$ is written in _____ notation.

4. A compound inequality containing the word "and" can be combined into a(n) _____ inequality.

5. The expression $(-1, 2]$ is written in _____ notation.

6. For any two sets A and B, the _____ of A and B, denoted _____, is defined as $\{x \mid x$ is an element of A *and* an element of $B\}$.

Basic Concepts

Exercises 1-2: Refer to Example 1 on page 541 in your text and the Section 8.3 lecture video.

Determine whether the given x-values are solutions to the compound inequalities.

1. $x+3<-1$ or $3x+2>4$ $x=-2, 2$

1. _____

   _____

2. $2x+1\leq 9$ and $x+2>-4$ $x=-5, 5$

2. _____

   _____

Symbolic Solutions and Number Lines

Exercises 3-8: Refer to Examples 2-5 on pages 542-545 in your text and the Section 8.3 lecture video.

3. Solve $4x-3<5$ and $3x+2\geq 2x+1$. Graph the solution set.

3. _____

Solve each inequality. Write the solution set in set-builder notation.

4. $-2<3x-5\leq 1$

4. _____

5. $-3<-x+3<5$

5. _____

6. $-\dfrac{3}{2}\leq\dfrac{1-m}{2}<3$

6. _____

7. If the ground-level temperature is $70°F$, the air temperature x miles above Earth's surface is cooler and can be modeled by $T(x) = 70 - 19x$. Find the altitudes at which the air temperature ranges from $44°F$ down to $30°F$. Round temperatures to the nearest tenth of a degree.

7. _____

8. Solve $2x - 3 \geq -4$ or $1 - 4x > -1$. Graph the solution set.

8. _____

Numerical and Graphical Solutions

Exercise 9: Refer to Example 6 on page 545 in your text and the Section 8.3 lecture video.

9. Medicare costs in billions of dollars may be modeled by $f(x) = 18x - 35,750$, where $1995 \leq x \leq 2007$. Estimate the years when Medicare costs were from \$160 to \$250 billion.

9. _____

Interval Notation

Exercise 10-15: Refer to Examples 7-8 on pages 546-547 in your text and the Section 8.3 lecture video.

Write each expression in interval notation.

10. $-3 < x \leq 4$

10. _____

11. $x \geq -\dfrac{1}{2}$

11. _____

12. $x < -3$ or $x \geq 1$ 12. _____

13. $\{x \mid x < 0$ and $x \geq -5\}$ 13. _____

14. $\left\{x \mid x \leq -2 \text{ or } x > \dfrac{5}{2}\right\}$ 14. _____

15.   Solve $\dfrac{1}{2}(6 - x) \leq 3$ and $3x - 1 > 5$. Write the solution set in   15. _____
in interval notation.

Chapter 8 Introduction to Functions
8.4 Other Functions and Their Properties

Expressing Domain and Range in Interval Notation ~ Absolute Value Function ~ Polynomial Functions ~ Rational Functions (Optional) ~ Operations on Functions

---

STUDY PLAN

Read: Read Section 8.4 on pages 551-562 in your textbook or eText.

Practice: Do your assigned exercises in your ☐ Book ☐ MyMathLab ☐ Worksheets

Review: Keep your corrected assignments in an organized notebook and use them to review for the test.

Key Terms
Exercises 1-5: Use the vocabulary terms listed below to complete each statement.
Note that some terms or expressions may not be used.

linear function   quadratic function
cubic function   absolute value function
rational function   polynomial function of one variable

1. We define a function called the _____ as $f(x) = |x|$.

2. A function having degree 1 is a(n) _____.

3. A function having degree 2 is a(n) _____.

4. A function having degree 3 is a(n) _____.

5. Let $p(x)$ and $q(x)$ be polynomials. Then a(n) _____ is given by
    $f(x) = \dfrac{p(x)}{q(x)}$. The domain of f includes all x-values such that $q(x) \neq 0$.

Expressing Domain and Range in Interval Notation

Exercises 1-4: Refer to Examples 1-2 on page 552 in your text and the Section 8.4 lecture video.

Write the domain for each function in interval notation.

1. $f(x) = 3x$

1. _____

2. $g(t) = \sqrt{t-4}$

2. _____

3. $h(x) = \dfrac{1}{5x}$

3. _____

4. Use the graph of f in the figure to write its domain and

4. _____

range in interval notation.

_____

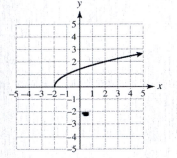

Polynomial Functions

Exercises 5-12: Refer to Examples 3-6 on pages 553-555 in your text and the Section 8.4 lecture video.

Determine whether f(x) represents a polynomial function. If possible, identify the type of polynomial function and its degree.

5. $f(x) = 3x^2 - 7x + 5$

5. _____

6. $f(x) = -\dfrac{5}{x+2}$

6. _____

7. $f(x) = 4x^{-0.5}$

7. _____

8. $f(x) = -1 + 4x$

8. _____

9. A graph of $f(x) = 4x - 4x^2 - 3x^3$ in shown in the figure, where $y = f(x)$. Evaluate $f(1)$ graphically and check your result symbolically.

9. _____

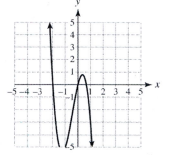

Evaluate f(x) at the given value of x.

10. $f(x) = -4x^3 + 2, \; x = -1$

10. _____

11. $f(x) = -2x^4 + x^2 - x, \; x = 0$

11. _____

12.   Let $f(t) = 1.96t^2 - 30t + 180$ model an athlete's heart rate (or pulse P) in beats per minute (bpm) t minutes after strenuous exercise has stopped, where $0 \le t \le 8$.

    (a)   What is the initial heart rate when the athlete stops exercising?

        12.(a)_____

    (b)   What is the heart rate after 5 minutes?

        (b)_____

Rational Functions (Optional)

Exercises 13-18: Refer to Examples 7-10 on pages 556-559 in your text and the Section 8.4 lecture video.

Write the domain of each function in interval notation.

13. $f(x) = \dfrac{x^2 + 4}{x + 7}$

    13. _____

14. $f(t) = \dfrac{4t}{t^2 - 5t + 4}$

    14. _____

15. $f(b) = \dfrac{-5}{b^3 + 2b^2}$

    15. _____

16.   Graph $f(x) = \dfrac{1}{x + 3}$. State the domain of f.

    16. _____

17. Use the table, the formula for $f(x)$, and the figure to evaluate $f(-2)$, $f(-1)$, and $f(1)$.

(a)

x	-3	-2	-1	0	1	2	3
$f(x)$	-3	___	1	0	$-\dfrac{1}{3}$	$-\dfrac{1}{2}$	$-\dfrac{3}{5}$

17.(a)_____

(b) $f(x) = -\dfrac{x}{x+2}$

(b)_____

(c)

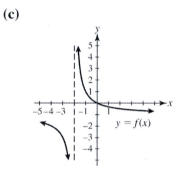

$y = f(x)$

(c)_____

18. The rational function given by $f(x) = \dfrac{100}{101-x}$, $5 \le x \le 100$, models participation inequality in a social network. In this formula, $f(x)$ outputs the percentage of the postings done by the least active x percent of the population.

(a) Evaluate $f(85)$. Interpret your answer.

18.(a)_____

(b) Solve the rational equation $f(x) = \dfrac{100}{101-x} = 10$. Interpret your answer.

(b)_____

Operations on Functions

Exercises 19-26: Refer to Examples 11-12 on page 561 in your text and the Section 8.4 lecture video.

Use $f(x) = 3x + 5$ and $g(x) = x^2$ to evaluate each of the following.

19. $(f + g)(-1)$

19. _____

20. $(fg)(-3)$

20. _____

21. $\left(\dfrac{f}{g}\right)(0)$

21. _____

22. $(f/g)(-2)$

22. _____

Use $f(x) = 4x - 1$ and $g(x) = x + 7$ to evaluate each of the following.

23. $(f + g)(x)$

23. _____

24. $(f - g)(x)$

24. _____

25. $(fg)(x)$

25. _____

26. $\left(\dfrac{f}{g}\right)(x)$

26. _____

Chapter 8 Introduction to Functions
8.5 Absolute Value Equations and Inequalities

Absolute Value Equations ~ Absolute Value Inequalities

STUDY PLAN

Read: Read Section 8.5 on pages 568-574 in your textbook or eText.

Practice: Do your assigned exercises in your ☐ Book ☐ MyMathLab ☐ Worksheets

Review: Keep your corrected assignments in an organized notebook and use them to review for the test.

Key Terms
Exercises 1-4: Use the expressions listed below to complete each statement.
Note that some expressions may not be used.

$$|ax + b| = |cx + d|$$
$$|ax + b| = k$$
$$|ax + b| > k$$
$$|ax + b| < k$$

1. If $k > 0$, then _____ is equivalent to $ax + b = k$ or $ax + b = -k$.

2. Let a, b, c, and d be constants. Then _____ is equivalent to $ax + b = cx + d$ or $ax + b = -(cx + d)$.

3. Let the solutions to $|ax + b| = k$ be c and d, where $c < d$ and $k > 0$. Then _____ is equivalent to $c < x < d$.

4. Let the solutions to $|ax + b| = k$ be c and d, where $c < d$ and $k > 0$. Then _____ is equivalent to $x < c$ or $x > d$.

Absolute Value Equations

Exercises 1-8: Refer to Examples 1-5 on pages 569-571 in your text and the Section 8.5 lecture video.

Solve each equation.

1. $|x| = 0$

1. _____

2. $|-4x| = 12$

2. _____

3. Solve $|3x - 1| = 14$ symbolically.

3. _____

Solve.

4. $|-2x + 1| + 2 = 4$

4. _____

5. $\left|\dfrac{1}{2}(x + 4)\right| = \dfrac{1}{4}$

5. _____

6. $|3 + 2x| = -4$

6. _____

7. $|1 - 4x| = 0$

7. _____

8. $|3x| = |2 - x|$

8. _____

Absolute Value Inequalities

Exercises 9-16: Refer to Examples 6-10 on pages 572-574 in your text and the Section 8.5 lecture video.

Solve each absolute value equation or inequality. Write the solution set for each inequality in interval notation.

9. $|1-2x|=5$ 9. _____

10. $|1-2x|\leq 5$ 10. _____

11. $|1-2x|>5$ 11. _____

12. $\left|\dfrac{2x+3}{5}\right|\geq 3$ 12. _____

13. A contractor is installing a custom-built window. The width w of the window is to be 32.4 inches and must be accurate to within 0.03 inches. Write an absolute value inequality that gives acceptable values for w.   13. _____

14. The inequality $|T-70|\leq 27$ models the range for the monthly average temperatures T in degrees Fahrenheit in Houston, Texas. Solve the inequality and interpret the results.   14. _____

Solve if possible.

15. $|t+2|+6\leq 5$ 15. _____

16. $|-2x+6|\geq -8$ 16. _____

Understanding Concepts through Multiple Approaches
(For additional practice, visit MyMathLab.)

17. Solve the absolute value inequality $|2x + 3| \geq 3$.

(a) Solve algebraically.

(b) Solve numerically using the table shown.

x	−4	−3	−2	−1	0	1
$\|2x + 3\|$						

(c) Solve visually.

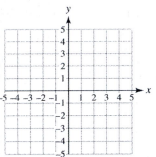

Did you get the same result using each method? Which method do you prefer? Explain why.

Chapter 9 Systems of Linear Equations
9.1 Systems of Linear Equations in Three Variables

Basic Concepts ~ Solving Linear Systems with Substitution and Elimination ~ Modeling Data ~ Systems of Equations with No Solutions ~ Systems of Equations with Infinitely Many Solutions

---

STUDY PLAN

Read: Read Section 9.1 on pages 591-600 in your textbook or eText.

Practice: Do your assigned exercises in your ☐ Book ☐ MyMathLab ☐ Worksheets

Review: Keep your corrected assignments in an organized notebook and use them to review for the test.

Key Terms
Exercises 1-2: Use the vocabulary terms listed below to complete each statement. Note that some terms or expressions may not be used. Some terms may be used more than once.

solve	solution
substitute	ordered triple
eliminate	linear system in three variables

1. When solving a(n) _____, we often use the variables x, y, and z. A solution is expressed as a(n) _____ (x, y, z).

2. Solving a Linear Systems in Three Variables
   Step 1: _____ one variable, such as x, from two of the equations.
   Step 2: Use the two resulting equations in two variables to _____ one of the variables, such as y. Solve for the remaining variable.
   Step 3: _____ z in one of the two equations from Step 2. _____ for the unknown variable y.
   Step 4: _____ values for y and z in one of the given equations and find x. The _____ is (x, y, z).

Basic Concepts

Exercises 1-3: Refer to Examples 1-3 on pages 592-593 in your text and the Section 9.1 lecture video.

1. Determine whether $(1,0,-3)$ or $(-2,1,2)$ is a solution to the system.   1. _____

$$-x + 2y - z = 2$$
$$3x - y + 2z = -3$$
$$2x + 2y + z = 0$$

2. The measure of the largest angle in a triangle is $20°$ greater than the   2. _____
    sum of the two smaller angles and $70°$ more than the smallest angle.
    Set up a system of three linear equations in three variables whose
    solution gives the measure of each angle.

3. The table shows the total cost of purchasing combinations of three   3. _____
    differently priced products, A, B, and C. Let a be the cost of prod-
    uct A, b be the price of product B, and c be the cost of product C.
    Write a system of linear equations whose solution gives appropriate
    values for a, b, and c.

A	B	C	Total Cost
2	1	1	$39
2	3	0	$49
1	1	4	$67

Solving Linear Systems with Substitution and Elimination

Exercises 4-6: Refer to Examples 4-6 on pages 594-596 in your text and the Section 9.1 lecture video.

4. Solve the following system.   4. _____

$$x - 3y + z = 4$$
$$2y + z = 1$$
$$z = 3$$

5. Solve the following system.   5. _____

$$x + 2y - 3z = -3$$
$$3x - 2y + z = 15$$
$$2x + 3y + 4z = -3$$

6. Seven hundred fifty tickets were sold for a concert, which   6. _____
    generated $4800 in revenue. The prices of the tickets were $4   _____
    for children, $5 for students, and $8 for adults. There were   _____
    200 fewer student tickets sold than adult tickets. Find the
    number or each type of ticket sold.

Modeling Data

Exercise 7: Refer to Example 7 on page 597 in your text and the Section 9.1 lecture video.

7. Solve the following linear system for a, b, and c.   7. _____

$$a + 14b + 3c = 160$$
$$a + 4b + 2c = 140$$
$$a + 10b + c = 40$$

Systems of Equations with No Solutions

Exercise 8: Refer to Example 8 on page 598 in your text and the Section 9.1 lecture video.

8. Solve the system, if possible.   8. _____

$$x - y + z = -3$$
$$-x - y + z = 2$$
$$y - z = -1$$

Systems of Equations with Infinitely Many Solutions

Exercise 9: Refer to Example 9 on pages 598-599 in your text and the Section 9.1 lecture video.

9. Solve the system.   9. _____

$$x + y - 2z = 0$$
$$2x + 3y - 4z = -1$$
$$x \qquad - 2z = 1$$

Chapter 9 Systems of Linear Equations
9.2 Matrix Solutions of Linear Equations

Representing Systems of Linear Equations with Matrices ~ Matrices and Social Networks ~ Gauss-Jordan Elimination ~ Using Technology to Solve Systems of Linear Equations (Optional)

STUDY PLAN

Read: Read Section 9.2 on pages 604-612 in your textbook or eText.

Practice: Do your assigned exercises in your   ☐ Book   ☐ MyMathLab   ☐ Worksheets

Review: Keep your corrected assignments in an organized notebook and use them to review for the test.

Key Terms
Exercises 1-7: Use the vocabulary terms listed below to complete each statement.
Note that some terms or expressions may not be used.

matrix	element
square	dimension
augmented	matrix row transformations
main diagonal	reduced row-echelon form
Gauss-Jordan elimination	

1. The _____ of a matrix is $m \times n$ if it has m rows and n columns.

2. A(n) _____ is a rectangular array of numbers.

3. A numerical method called _____ is used to solve a linear system. It makes use of _____.

4. If the number of rows equals the number of columns, the matrix is a(n) _____ matrix.

5. A convenient matrix form for representing a system of linear equations is _____.

6. The numbers $1, -3, 5$ are elements of the _____ of the _____

   matrix $\begin{bmatrix} 1 & 2 & 2 & | & 7 \\ 0 & -3 & 1 & | & -1 \\ 2 & 4 & 5 & | & 0 \end{bmatrix}$.

7. Each number in a matrix is called a(n) _____.

Representing Systems of Linear Equations with Matrices

Exercises 1-2: Refer to Example 1 on page 605 in your text and the Section 9.2 lecture video.

Represent each linear system with an augmented matrix. State the dimension of the matrix.

1. $x + 3y = 13$
   $5x - 3y = -25$

   1. _____

2. $x - y + 2z = -2$
   $2x - 4y + 3z = 1$
   $-2x + 2y - 3z = 2$

   2. _____

Matrices and Social Networks

Exercise 3: Refer to Example 2 on page 605 in your text and the Section 9.2 lecture video.

3. Use a matrix to model the social network shown in the figure.

   3. _____

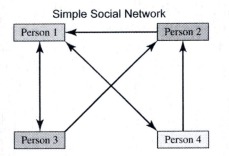

Simple Social Network

Gauss-Jordan Elimination

Exercises 4-8: Refer to Examples 3-6 on pages 606-609 in your text and the Section 9.2 lecture video.

Each matrix represents a system of linear equations. Find the solution.

4. $\begin{bmatrix} 1 & 0 & | & 0 \\ 0 & 1 & | & -5 \end{bmatrix}$

   4. _____

5. $\begin{bmatrix} 1 & 0 & 0 & | & -2 \\ 0 & 1 & 0 & | & 4 \\ 0 & 0 & 1 & | & -5 \end{bmatrix}$

   5. _____

Use Gauss-Jordan elimination to transform the augmented matrix of the linear system into reduced row-echelon form. Find the solution.

6.
$$x - 3y = -4$$
$$2x - y = -3$$

6. _____

7.
$$x + 2y + z = -5$$
$$x \quad\quad - 2z = 4$$
$$2x + 3y \quad\quad = -5$$

7. _____

8. A total of $15,000 is invested in three funds that grew at a rate of 4%, 7%, and 8% over 1 year. After 1 year, the combined value of the three funds had grown by $1040. Twice as much money is invested at 8% as at 7%. Find the amount invested in each fund.

8. _____
_____
_____

Using Technology to Solve Systems of Linear Equations (Optional)

Exercises 9-11: Refer to Examples 7-8 on pages 610-611 in your text and the Section 9.2 lecture video.

Use a graphing calculator to solve the following systems of equations.

9.
$$x + y = 0$$
$$4x - 2y = 3$$

9. _____

10.
$$x + y + z = -1$$
$$-2x + y - 3z = -1$$
$$3x - 2y + z = -9$$

10. _____

11. The selling price of a home depends on several factors, such as size and age. The accompanying table shows the selling price for each of three homes. In this table, price P is given in thousands of dollars, age A in years, and home size S in thousands of square feet. The data can be modeled by the equation $P = a + bA + cS$.

Price (P)	Age (A)	Size (S)
$215	15	2
$450	10	4
$ 75	25	1

(a) Set up a system of equations whose solution gives values for constants a, b, and c.

11.(a)_____

(b) Solve the system.

(b)_____

(c) Predict the selling price of a home with $A = 5$ years and $S = 3$ thousand square feet.

(c)_____

Chapter 9 Systems of Linear Equations
9.3 Determinants

Calculation of Determinants ~ Area of Regions ~ Cramer's Rule

---

STUDY PLAN

Read: Read Section 9.3 on pages 616-620 in your textbook or eText.

Practice: Do your assigned exercises in your ☐ Book ☐ MyMathLab ☐ Worksheets

Review: Keep your corrected assignments in an organized notebook and use them to review for the test.

Key Terms

Exercises 1-4: Use the vocabulary terms listed below to complete each statement. Note that some terms or expressions may not be used. Some terms may be used more than once.

> **minor**
> **determinant**
> **Cramer's rule**
> **expansion by minors**
> $a,\ b,\ c,\ d,\ a_1,\ a_2,\ a_3,\ a_4,\ b_1,\ b_2,\ b_3,\ b_4,\ c_1,\ c_2,\ c_3,\ c_4$

1. The _____ of $A = \begin{bmatrix} a & - \\ c & d \end{bmatrix}$ is a real number defined by $\det A = ad - cb$.

2. We can use _____(s) of 2×2 matrices to find _____(s) of 3×3 matrices. This method is called _____.

3. $\det A = \det \begin{bmatrix} a_1 & b_1 & c_1 \\ a_2 & b_2 & c_2 \\ a_3 & b_3 & c_3 \end{bmatrix} = a_1 \cdot \det \begin{bmatrix} b_2 & c_2 \\ - & c_3 \end{bmatrix} - \_ \cdot \det \begin{bmatrix} b_1 & - \\ b_3 & c_3 \end{bmatrix} + a_3 \cdot \det \begin{bmatrix} b_1 & c_1 \\ b_2 & - \end{bmatrix}$

    The 2×2 matrices in this equation are called _____(s).

4. The solution to the system of linear equations $\begin{array}{l} a_1 x + b_1 y = c_1 \\ a_2 x + b_2 y = c_2 \end{array}$ is given by $x = \frac{E}{D}$ and $y = \frac{F}{D}$,

    where $E = \det \begin{bmatrix} - & b_1 \\ c_2 & b_2 \end{bmatrix}$, $F = \det \begin{bmatrix} a_1 & c_1 \\ a_2 & - \end{bmatrix}$, $D = \det \begin{bmatrix} a_1 & b_1 \\ - & b_2 \end{bmatrix} \neq 0$. This method is called

    _____.

Calculation of Determinants

Exercises 1-6: Refer to Examples 1-3 on pages 616-618 in your text and the Section 9.3 lecture video.

Evaluate det *A by hand for each* 2×2 *matrix.*

1. $A = \begin{bmatrix} 2 & -1 \\ 3 & -4 \end{bmatrix}$ 1. _____

2. $A = \begin{bmatrix} -3 & -4 \\ -5 & 1 \end{bmatrix}$ 2. _____

Evaluate det *A.*

3. $A = \begin{bmatrix} 2 & -2 & 0 \\ 1 & -3 & 4 \\ 0 & 1 & 7 \end{bmatrix}$ 3. _____

4. $A = \begin{bmatrix} 1 & -2 & 0 \\ 3 & -5 & -2 \\ 1 & 1 & 4 \end{bmatrix}$ 4. _____

Find each determinant of A, using a graphing calculator.

5. $A = \begin{bmatrix} 2 & -2 & 0 \\ 1 & -3 & 4 \\ 0 & 1 & 7 \end{bmatrix}$ 5. _____

6. $A = \begin{bmatrix} 1 & -2 & 0 \\ 3 & -5 & -2 \\ 1 & 1 & 4 \end{bmatrix}$ 6. _____

Area of Regions

Exercise 7: Refer to Example 4 on page 618 in your text and the Section 9.3 lecture video.

7. A triangular parcel of land is shown in the figure. If all units are miles, find the area of the parcel of land by using a determinant.

7. _____

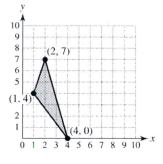

Cramer's Rule

Exercises 8-9: Refer to Example 5 on page 619 in your text and the Section 9.3 lecture video.

Use Cramer's rule to solve the following linear systems.

8. $2x + 3y = 0$
   $5x - 2y = 19$

8. _____

9. $5x - 3y = -7$
   $-4x + 2y = 9$

9. _____

Name: _____ Course/Section: _____ Instructor: _____

Chapter 10 Radical Expressions and Functions
10.1 Radical Expressions and Functions

Radical Notation ~ The Square Root Function ~ The Cube Root Function

---

STUDY PLAN

Read: Read Section 10.1 on pages 632-640 in your textbook or eText.

Practice: Do your assigned exercises in your ☐ Book ☐ MyMathLab ☐ Worksheets

Review: Keep your corrected assignments in an organized notebook and use them to review for the test.

Key Terms
Exercises 1-8: Use the vocabulary terms listed below to complete each statement.
Note that some terms or expressions may not be used.

radicand   *n*th root
cube root   index
radical sign   even root
odd root   square root
radical expression   principal *n*th root

1. The symbol $\sqrt{}$ is called the _____.

2. The _____ function is given by $f(x) = \sqrt[3]{x}$.

3. If $a > 0$ and n is even, then the positive nth root is denoted $\sqrt[n]{a}$ and is called the _____ of a.

4. An expression containing a radical sign is called a(n) _____.

5. The _____ function is given by $f(x) = \sqrt{x}$.

6. The number b is a(n) _____ of a if $b^n = a$.

7. The equation $\sqrt[n]{a} = b$ means that $b^n = a$, where n is a natural number called the _____. If n is odd, we are finding a(n) _____ and if n is even, we are finding a(n) _____

8. The expression under the radical sign is called the _____.

Radical Notation

Exercises 1-19: Refer to Examples 1-7 on pages 632-636 in your text and the Section 10.1 lecture video.

1. Find the square roots of 144.   1. _____

Evaluate each square root.

2. $\sqrt{64}$ 2. _____

3. $\sqrt{0.36}$ 3. _____

4. $\sqrt{\dfrac{16}{25}}$ 4. _____

5. $\sqrt{d^2}$, $d < 0$ 5. _____

6. Approximate $\sqrt{19}$ to the nearest thousandth.   6. _____

7. A formula for calculating the distance d in miles that one can see
    to the horizon on a clear day is approximated by $d = 1.22\sqrt{x}$,
    where x is the elevation, in feet, of person.

    (a) Approximate how far a 5.75-foot-tall person can see to the   7.(a)_____
        horizon.

    (b) Approximate how far a person can see from a 9500-foot   (b)_____
        mountain.

Evaluate the cube root. Approximate your answer to the nearest hundredth when appropriate.

8. $\sqrt[3]{27}$ 8. _____

9. $\sqrt[3]{-64}$ 9. _____

10. $\sqrt[3]{\dfrac{1}{8}}$ 10. _____

11. $\sqrt[3]{a^{12}}$ 11. _____

12. $\sqrt[3]{12}$ 12. _____

Evaluate each root, if possible.

13. $\sqrt[4]{625}$ 13. _____

14. $\sqrt[5]{-243}$ 14. _____

15. $\sqrt[4]{-16}$ 15. _____

16. $-\sqrt[4]{16}$ 16. _____

Write each expression in terms of an absolute value.

17. $\sqrt{z^2}$

17. _____

18. $\sqrt{(x-2)^2}$

18. _____

19. $\sqrt{x^2+6x+9}$

19. _____

The Square Root Function

Exercises 20-26: Refer to Examples 8-12 on pages 636-639 in your text and the Section 10.1 lecture video.

If possible, evaluate $f(1)$ and $f(-2)$ for each $f(x)$.

20. $f(x)=\sqrt{4x-3}$

20. _____

_____

21. $f(x)=\sqrt{13-x^2}$

21. _____

_____

22. If a football is kicked x feet high, then the time T in seconds that the ball is in the air is given by the function

$$T(x)=\frac{\sqrt{x}}{2}.$$

(a) Find the hang time if the ball is kicked 30 feet into the air.

22.(a)_____

(b) Does the hang time triple if the ball is kicked 90 feet into the air?

(b)_____

23. Let $f(x) = \sqrt{x-3}$.

 (a) Find the domain of f. Write your answer in interval notation.   23.(a)_____

 (b) Graph $y = f(x)$ and compare it to the graph of $y = \sqrt{x}$.   (b)_____

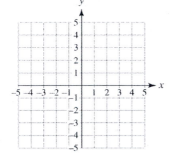

Find the domain of each function. Write your answer in interval notation.

24. $g(x) = \sqrt{2x+4}$ 15. _____

25. $f(x) = \sqrt{x^2+3}$ 16. _____

26. The function $R(x) = 132\sqrt{x}$ gives the total revenue per year in thousands of dollars generated by a small business having x employees. A graph of $y = R(x)$ is shown in the figure.

Revenue as a Function of Employees

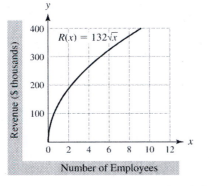

 (a) Approximate values for $R(4)$, $R(8)$, and $R(12)$ by using   26.(a)_____
 both the formula and the graph.

 (b) Evaluate $R(12) - R(8)$ and $R(8) - R(4)$ using the formula.   (b)_____
 Interpret your answer.

The Cube Root Function

Exercises 27-28: Refer to Example 13 on page 640 in your text and the Section 10.1 lecture video.

Evaluate $f(1)$ and $f(-3)$ for each $f(x)$.

27. $f(x) = \sqrt[3]{2x^2 - 1}$

27. _____

_____

28. $f(x) = \sqrt[3]{9 - x^2}$

28. _____

_____

Chapter 10 Radical Expressions and Functions
10.2 Rational Exponents

Basic Concepts ~ Properties of Rational Exponents

STUDY PLAN

Read: Read Section 10.2 on pages 643-649 in your textbook or eText.

Practice: Do your assigned exercises in your ☐ Book ☐ MyMathLab ☐ Worksheets

Review: Keep your corrected assignments in an organized notebook and use them to review for the test.

Key Terms
Exercises 1-10: Use the expressions listed below to complete each statement.
Note that some expressions may not be used.

$$\frac{1}{a^p} \qquad \left(\frac{b}{a}\right)^p \quad \frac{1}{a^{m/n}} = \frac{1}{\sqrt[n]{a^m}} \qquad \sqrt[n]{a^m} = \left(\sqrt[n]{a}\right)^m \qquad a^{p+q}$$

$$\sqrt[n]{a} \qquad\qquad a^{pq} \qquad\qquad\qquad\qquad\qquad\qquad\qquad a^p$$

$$|a| \qquad \left(\frac{a}{b}\right)^p \qquad\qquad \frac{a^p}{b^p}$$

$$\qquad\qquad\qquad\qquad\qquad\qquad a^{p-q}$$

1. If n is an integer greater than 1 and a is a real number, then $a^{1/n} =$ _____.

2. If m and n are positive integers with $\frac{m}{n}$ in lowest terms, then $a^{m/n} =$ _____.

3. If m and n are positive integers with $\frac{m}{n}$ in lowest terms, then $a^{-m/n} =$ _____, $a \neq 0$.

4. $a^p \cdot a^q =$ _____

5. $a^{-p} =$ _____

6. $\left(\frac{a}{b}\right)^{-p} =$ _____

7. $\frac{a^p}{a^q} =$ _____

8. $\left(a^p\right)^q =$ _____

9. $(ab)^p =$ _____

10. $\left(\frac{a}{b}\right)^p =$ _____

Basic Concepts

Exercises 1-14: Refer to Examples 1-6 on pages 644-647 in your text and the Section 10.2 lecture video.

Write each expression in radical notation. Then evaluate the expression and round to the nearest hundredth when appropriate.

1. $49^{\frac{1}{2}}$ 1. _____

2. $15^{\frac{1}{3}}$ 2. _____

3. $x^{\frac{1}{4}}$ 3. _____

4. $(3z)^{\frac{1}{2}}$ 4. _____

Write each expression in radical notation. Evaluate the expression by hand when possible.

5. $(-64)^{\frac{2}{3}}$ 5. _____

6. $32^{\frac{3}{5}}$ 6. _____

7. The function A, given by $A(x) = 7.3x^{\frac{7}{16}}$, computes the percen-   7. _____
    tage of viewers who abandon an online video after x seconds.   _____
    Find $A(20)$ and $A(60)$. Interpret your answers.

Write each expression in radical notation and then evaluate.

8. $16^{-3/4}$ 8. _____

9. $-27^{-2/3}$ 9. _____

Use rational exponents to write each radical expression.

10. $\sqrt[6]{x^5}$ 10. _____

11. $\dfrac{1}{\sqrt{a^7}}$ 11. _____

12. $\sqrt[4]{(x+4)^3}$ 12. _____

13. $\sqrt[4]{s^4 - t^4}$ 13. _____

14. When smaller (four-legged) animals walk, they tend to take faster, shorter steps, whereas larger animals tend to take slower, longer steps. If an animal is h feet high at the shoulder, then the number N of steps per second that the animal takes while walking can be estimated by $N(h) = 1.6h^{-1/2}$. Estimate the stepping frequency for Gibson, one of the tallest dogs in the world, measuring 42 inches at the shoulder.

    14. _____

Properties of Rational Exponents

Exercises 15-22: Refer to Examples 7-8 on page 648 in your text and the Section 10.2 lecture video.

Write each expression using rational exponents and simplify. Write the answer with a positive exponent. Assume that all variables are positive numbers.

15. $\sqrt{x} \cdot \sqrt[4]{x}$ 15. _____

16. $\sqrt[4]{32x^6}$ 16. _____

17. $\dfrac{\sqrt[3]{27x}}{\sqrt{x}}$ 17. _____

18. $\left(\dfrac{49}{y^2}\right)^{-1/2}$ 18. _____

Write each expression with positive rational exponents and simplify, if possible.

19. $\sqrt{\sqrt[4]{z+3}}$ 19. _____

20. $\sqrt[4]{y^{12}}$ 20. _____

21. $\dfrac{a^{-1/3}}{b^{-1/4}}$ 21. _____

22. $\sqrt{x}\left(1+\sqrt{x}\right)$ 22. _____

Chapter 10 Radical Expressions and Functions
10.3 Simplifying Radical Expressions

Product Rule for Radical Expressions ~ Quotient Rule for Radical Expressions

---

STUDY PLAN

Read: Read Section 10.3 on pages 652-659 in your textbook or eText.

Practice: Do your assigned exercises in your   ☐ Book ☐ MyMathLab ☐ Worksheets

Review: Keep your corrected assignments in an organized notebook and use them to review for the test.

Key Terms
Exercises 1-5: Use the vocabulary terms and expressions listed below to complete each statement. Note that some terms or expressions may not be used.

perfect cube
perfect square
perfect *n*th power

$\sqrt[n]{a \cdot b}$

$\dfrac{\sqrt[n]{a}}{\sqrt[n]{b}}$

1. An integer a is a(n) _____ if there exists an integer b such that $b^n = a$.

2. Let a and b be real numbers, where $\sqrt[n]{a}$ and $\sqrt[n]{b}$ are both defined and $b \neq 0$. Then $\sqrt[n]{\dfrac{a}{b}} =$ _____.

3. An integer a is a(n) _____ if there exists an integer b such that $b^3 = a$.

4. Let a and b be real numbers, where $\sqrt[n]{a}$ and $\sqrt[n]{b}$ are both defined. Then $\sqrt[n]{a} \cdot \sqrt[n]{b} =$ _____.

5. An integer a is a(n) _____ if there exists an integer b such that $b^2 = a$.

Product Rule for Radical Expressions

Exercises 1-19: *Refer to Examples 1-6 on pages 653-656 in your text and the Section 10.3 lecture video.*

Multiply each radical expression.

1. $\sqrt{2} \cdot \sqrt{50}$

1. _____

2. $\sqrt[3]{-5} \cdot \sqrt[3]{25}$

2. _____

3. $\sqrt[4]{\dfrac{1}{2}} \cdot \sqrt[4]{\dfrac{1}{4}} \cdot \sqrt[4]{\dfrac{1}{2}}$

3. _____

Multiply each radical expression. Assume that all variables are positive.

4. $\sqrt{x^3} \cdot \sqrt{x^5}$

4. _____

5. $\sqrt[3]{4y} \cdot \sqrt[3]{7y}$

5. _____

6. $\sqrt{6a} \cdot \sqrt{b}$

6. _____

7. $\sqrt[5]{\dfrac{9s}{t}} \cdot \sqrt[5]{\dfrac{27t}{s}}$

7. _____

Simplify each expression.

8. $\sqrt{75}$

8. _____

9. $\sqrt[3]{81}$

9. _____

10. $\sqrt{24}$

10. _____

11. $\sqrt[4]{162}$

11. _____

12. A circular curve without any banking has a radius of 900 feet. The formula $L = \sqrt{2.25R}$ calculates the speed limit L in miles per hour for a curve with radius R feet.

(a) Simplify this formula.

12. (a)_____

(b) Determine the speed limit for the curve.

(b)_____

Simplify each expression. Assume that all variables are positive.

13. $\sqrt{16y^6}$

13. _____

14. $\sqrt{8x^5}$

14. _____

15. $\sqrt[3]{-54a^4b^6}$

15. _____

16. $\sqrt[3]{16m^2} \cdot \sqrt[3]{4mn}$

16. _____

Simplify each expression. Write your answer in radical notation.

17. $\sqrt{6} \cdot \sqrt[3]{6}$ 17._____

18. $\sqrt{3} \cdot \sqrt[4]{9}$ 18._____

19. $\sqrt[4]{b} \cdot \sqrt[5]{b}$ 19._____

Quotient Rule for Radical Expressions

Exercises 20-29: Refer to Examples 7-10 on pages 656-658 in your text and the Section 10.3 lecture video.

Simplify each radical expression. Assume that all variables are positive.

20. $\sqrt[3]{\dfrac{3}{8}}$ 20._____

21. $\sqrt[4]{\dfrac{x}{81}}$ 21._____

22. $\sqrt{\dfrac{25}{t^2}}$ 22._____

23. $\dfrac{\sqrt{45}}{\sqrt{5}}$ 23._____

24. $\dfrac{\sqrt[3]{2}}{\sqrt[3]{250}}$ 24. _____

25. $\dfrac{\sqrt{m^3 n}}{\sqrt{mn}}$ 25. _____

26. $\sqrt[4]{\dfrac{y^5}{16x^4}}$ 26. _____

27. $\sqrt{\dfrac{3x}{5}} \cdot \sqrt{\dfrac{12x^3}{5}}$ 27. _____

Simplify each expression. Assume all radicands are positive.

28. $\sqrt{x+4} \cdot \sqrt{x-4}$ 28. _____

29. $\dfrac{\sqrt[3]{x^2 + 7x + 6}}{\sqrt[3]{x+6}}$ 29. _____

Chapter 10 Radical Expressions and Functions
10.4 Operations on Radical Expressions

Addition and Subtraction ~ Multiplication ~ Rationalizing the Denominator

---

STUDY PLAN

Read: Read Section 10.4 on pages 661-670 in your textbook or eText.

Practice: Do your assigned exercises in your ☐ Book ☐ MyMathLab ☐ Worksheets

Review: Keep your corrected assignments in an organized notebook and use them to review for the test.

Key Terms
Exercises 1-3: Use the vocabulary terms listed below to complete each statement.
Note that some terms or expressions may not be used.

conjugate
radicand
like radical
radical expression
rationalizing the denominator

1. _____(s) have the same index and the same radicand.

2. The process of removing any radical expressions from the denominator is called
_____.

3. The _____ of $\sqrt{a} - \sqrt{b}$ is $\sqrt{a} + \sqrt{b}$.

Addition and Subtraction

Exercises 1-24: *Refer to Examples 1-9 on pages 662-666 in your text and the Section 10.4 lecture video.*

If possible, add the expressions and simplify.

1. $5\sqrt{7} + 8\sqrt{7}$

1. _____

2. $4\sqrt[3]{5} + \sqrt[3]{5}$

2. _____

3. $3 + 2\sqrt{3}$

3. _____

4. $\sqrt{6} + \sqrt{10}$

4. _____

Write each pair of terms as like radicals, if possible.

5. $\sqrt{75}, \sqrt{48}$

5. _____

6. $\sqrt{8}, \sqrt{4}$

6. _____

7. $7\sqrt[3]{24}, 2\sqrt[3]{81}$

7. _____

Add the expressions and simplify.

8. $\sqrt{18} + 4\sqrt{2}$

8. _____

9. $\sqrt[3]{24} + \sqrt[3]{3}$

9. _____

10. $5\sqrt{2} + \sqrt{8} + \sqrt{18}$

10. _____

Add the expressions and simplify. Assume that all variables are positive.

11. $\sqrt[4]{32} + 3\sqrt[4]{2}$

11. _____

12. $-4\sqrt{9x} + \sqrt{x}$

12. _____

13. $2\sqrt{5b} + 8\sqrt{20b} + 3\sqrt{45b}$

13. _____

Simplify the expressions.

14. $6\sqrt{5} - 2\sqrt{5}$

14. _____

15. $4\sqrt[3]{7} - 3\sqrt[3]{7} + \sqrt[3]{10}$

15. _____

16. $6\sqrt{t} + \sqrt[3]{t} - 3\sqrt{t}$

16. _____

Subtract and simplify. Assume that all variables are positive.

17. $5\sqrt[3]{m^2 n} - \sqrt[3]{m^2 n}$

17._____

18. $\sqrt{9y^3} - \sqrt{y^3}$

18._____

19. $\sqrt[3]{\dfrac{7x}{8}} - \dfrac{\sqrt[3]{7x}}{4}$

19._____

20. The functions given by $N(x) = 650\sqrt{x} + 5000$ (new technology)
and $O(x) = 225\sqrt{x} + 1700$ (old technology) approximate the
increase in revenue resulting from investing x dollars in equip-
ment (per worker).

   (a) Find their difference $D(x) = N(x) - O(x)$. Simplify
   your answer.

20.(a)_____

   (b) Evaluate $D(40,000)$ and interpret the result.

(b)_____

Subtract and simplify. Assume that all variables are positive.

21. $\dfrac{7\sqrt{2}}{4} - \dfrac{2\sqrt{2}}{3}$

21._____

22. $\sqrt[4]{81x^6 y^9} - \sqrt[4]{16x^2 y}$

22._____

23. $2\sqrt[3]{\dfrac{t^4}{8}} - 4\sqrt[3]{t}$

23._____

24. Find the exact perimeter of a rectangle with length $\sqrt{75}$ feet   24. _____
    and width $\sqrt{12}$ feet. Then approximate your answer to the
    nearest hundredth of a foot.

Muliplication

Exercises 25-26: Refer to Example 10 on pages 666-667 in your text and the Section 10.4 lecture video.

Multiply and simplify.

25. $\left(\sqrt{a}+6\right)\left(\sqrt{a}-2\right)$ 25. _____

26. $\left(2-\sqrt{5}\right)\left(2+\sqrt{5}\right)$ 26. _____

Rationalizing the Denominator

Exercises 27-35: Refer to Examples 11-15 on pages 667-670 in your text and the Section 10.4 lecture video.

Rationalize each denominator. Assume that all variables are positive.

27. $\dfrac{1}{\sqrt{3}}$ 27. _____

28. $\dfrac{2}{7\sqrt{2}}$ 28. _____

29. $\sqrt{\dfrac{y}{50}}$ 29. _____

30. $\dfrac{x^2 y}{\sqrt{x^3}}$ 30. _____

31. An equilateral triangle has sides of length $\dfrac{x}{\sqrt{3}}$. Find the perimeter and rationalize the denominator.

31. _____

Rationalize the denominator.

32. $\dfrac{3}{\sqrt{5}+2}$

32. _____

33. $\dfrac{2-\sqrt{3}}{3+\sqrt{3}}$

33. _____

34. $\dfrac{\sqrt{x}}{\sqrt{x}-4}$

34. _____

35. Rationalize the denominator of $\dfrac{4}{\sqrt[3]{y}}$.

35. _____

Chapter 10 Radical Expressions and Functions
10.5 More Radical Functions

Root Functions ~ Power Functions ~ Modeling with Power Functions (Optional)

STUDY PLAN

Read: Read Section 10.5 on pages 673-677 in your textbook or eText.

Practice: Do your assigned exercises in your   ☐ Book ☐ MyMathLab ☐ Worksheets

Review: Keep your corrected assignments in an organized notebook and use them to review for the test.

Key Terms
Exercise 1: Use the vocabulary terms listed below to complete each statement.
Note that some terms or expressions may not be used.

   root function
   radical function
   power function
   rational function

1.   If a function f can be represented by $f(x) = x^p$, where p is a rational number, then f is

   a(n) _____. If $p = \dfrac{1}{n}$, where $n \geq 2$ is an integer, then f is also a(n)

   _____, which is given by $f(x) = \sqrt[n]{x}$.

Root Functions

Exercises 1-4: Refer to Example 1 on page 674 in your text and the Section 10.5 lecture video.

If possible, evaluate each root function f at the given x-values.

1. $f(x) = \sqrt{x+4}, \; x = -5, \; x = 5$ 1. _____

2. $f(x) = \sqrt[3]{3-x}, \; x = -5, \; x = 30$ 2. _____

3. $f(x) = \sqrt[4]{x-5}, \; x = 1, \; x = 21$ 3. _____

4. $f(x) = \sqrt[5]{x}, \; x = -243, \; x = 1$ 4. _____

Power Functions

Exercises 5-8: Refer to Examples 2-4 on pages 675-676 in your text and the Section 10.5 lecture video.

If possible, evaluate f(x) at the given value of x.

5. $g(x) = x^{0.75}$ at $x = -81$ 5. _____

6. $g(x) = x^{2/3}$ at $x = -27$ 6. _____

7.   The surface area of a person who is 70 inches tall and weighs w
     pounds can be estimated by $S(w) = 342w^{0.425}$, where S is in square
     inches.

     (a) Find S if this person weighs 190 pounds.   7. (a)_____

     (b) If the person gains 20 pounds, by how much does the   (b)_____
          person's surface area increase?

8.   The graphs of three power functions,   8. _____

     $$f(x) = x^{0.5}, \quad g(x) = x^{1.8}, \quad \text{and} \quad h(x) = x^{\frac{2}{3}},$$

     are shown below, where the two decimal exponents have
     been written as fractions. Discuss how the value of p
     affects the graph of $y = x^p$ when $x > 1$ and when $0 < x < 1$.

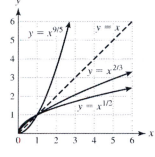

Modeling with Power Functions (Optional)

Exercise 9: Refer to Example 5 on pages 676-677 in your text and the Section 10.5 lecture video.

9.   Biologists have found that the weight W of a bird and the length
     L of its wing span are related by the equation $L = kW^{\frac{1}{3}}$, where
     k is constant. The following table lists L and W for one species
     of bird.

W (kilograms)	0.2	0.3	0.5	0.8
L (meters)	0.498	0.569	0.676	0.789

     (a) Use the data to approximate the value of k.   9. (a)_____

     (b) Find the wing span of a 0.6-kilogram bird.   (b)_____

Chapter 10 Radical Expressions and Functions
10.6 Equations Involving Radical Expressions

Solving Radical Equations ~ The Distance Formula ~ Solving the Equation $x^n = k$

STUDY PLAN

Read: Read Section 10.6 on pages 681-690 in your textbook or eText.

Practice: Do your assigned exercises in your   ☐ Book   ☐ MyMathLab   ☐ Worksheets

Review: Keep your corrected assignments in an organized notebook and use them to review for the test.

Key Terms
Exercises 1-3: Use the vocabulary terms listed below to complete each statement.
Note that some terms or expressions may not be used.

distance
extraneous solution
Pythagorean theorem

1. After applying the power rule, the new equation can have solutions that do not satisfy the given equation, which are sometimes called _____(s).

2. The _____ states that if a right triangle has legs a and b with hypotenuse c, then $a^2 + b^2 = c^2$.

3. The _____ between the points (x_1, y_1) and (x_2, y_2) in the xy-plane is
$$d = \sqrt{(x_2 - x_1)^2 + (y_2 - y_1)^2}.$$

Solving Radical Equations

Exercises 1-7: Refer to Examples 1-7 on pages 682-686 in your text and the Section 10.6 lecture video.

1. Solve $\sqrt{2x+5} = 5$. Check your solution.

1. _____

2. Solve $\sqrt{3x+7} + 3 = 10$.

2. _____

3. Solve $\sqrt{8x+9} = 3x-1$. Check your results and then solve the equation graphically

3. _____

4. Solve $\sqrt{5x} - 1 = \sqrt{3x+1}$.

4. _____

5. $\sqrt[3]{3x+4} = -2$

5. _____

6. Solve the equation $300 = 100\sqrt[3]{W^2}$ to estimate the weight in pounds of a bird having wings with an area of 300 square inches.

6. _____

7. Solve $x^{1/3} = 5 - x^2$ graphically.

7. _____

The Distance Formula

Exercises 8-9: Refer to Examples 8-9 on pages 686-688 in your text and the Section 10.6 lecture video.

8. A 20 foot ladder is positioned 12 feet from a wall. How far up the wall does the ladder reach?

8. _____

9. Find the distance between the points $(5,5)$ and $(-6,-6)$.

9. _____

Solving the Equation $x^n = k$

Exercises 10-13: Refer to Examples 10-11 on pages 688-689 in your text and the Section 10.6 lecture video.

Solve each equation.

10. $x^3 = -27$

10. _____

11. $x^2 = -2$

11. _____

12. $2(x-5)^4 = 162$

12. _____

13. The formula $W(v) = 2.8v^3$ is used to calculate the watts generated when there is a wind velocity of v miles per hour.

   (a) Find a function v that calculates the wind velocity when W watts are produced.

   13. (a)_____

   (b) Find the wind velocity when 1500 watts are produced.

   (b)_____

Understanding Concepts through Multiple Approaches
(For additional practice, visit MyMathLab.)

14. Solve the equation $2\sqrt{x+3}=12$.

(a) Solve algebraically.

(b) Solve numerically using the table feature of a graphing calculator.

(c) Solve visually using a graphing calculator.

Did you get the same result using each method? Which method do you prefer? Explain why.

Chapter 10 Radical Expressions and Functions
10.7 Complex Numbers

Basic Concepts ~ Addition, Subtraction, and Multiplication ~ Powers of i ~ Complex Conjugates and Division

---

STUDY PLAN

Read: Read Section 10.7 on pages 695-701 in your textbook or eText.

Practice: Do your assigned exercises in your   ☐ Book ☐ MyMathLab ☐ Worksheets

Review: Keep your corrected assignments in an organized notebook and use them to review for the test.

Key Terms
Exercises 1-6: Use the vocabulary terms listed below to complete each statement. Note that some terms or expressions may not be used. Some terms may be used more than once.

real	**imaginary**
complex	**pure imaginary**
standard form	**complex conjugate**

1. A(n) _____ number can be written in _____, as $a + bi$, where a and b are real numbers. The _____ part is a and the _____ part is b.

2. A complex number $a + bi$ with $b \neq 0$ is a(n) _____ number.

3. The _____ of $a + bi$ is $a - bi$.

4. A complex number $a + bi$ with $a = 0$ and $b \neq 0$ is sometimes called a(n) _____ number.

5. A number i, defined by $i = \sqrt{-1}$ or $i^2 = -1$, is called the _____ unit.

6. A complex number $a + bi$ with $b = 0$ is a(n) _____ number.

Basic Concepts

Exercises 1-3: Refer to Example 1 on page 696 in your text and the Section 10.7 lecture video.

Write each square root using the imaginary unit i.

1. $\sqrt{-16}$

1. _____

2. $\sqrt{-5}$

2. _____

3. $\sqrt{-50}$

3. _____

Addition, Subtraction, and Multiplication

Exercises 4-7: Refer to Examples 2-3 on pages 697-698 in your text and the Section 10.7 lecture video.

Write each sum or difference in standard form.

4. $(2+9i)+7$

4. _____

5. $(-4+3i)-(6+i)$

5. _____

Write each product in standard form.

6. $(2+i)(3+4i)$

6. _____

7. $(-2+6i)(-2-6i)$

7. _____

Powers of i

Exercises 8-10: Refer to Example 4 on page 699 in your text and the Section 10.7 lecture video.

Evaluate each expression.

8. i^8 8. _____

9. i^{13} 9. _____

10. i^{102} 10. _____

Complex Conjugates and Division

Exercises 11-16: Refer to Examples 5-6 on pages 699-700 in your text and the Section 10.7 lecture video.

Find the complex conjugate of each number.

11. $4+i$ 11. _____

12. $-5-2i$ 12. _____

13. $6i$ 13. _____

14. 3 14. _____

Write each quotient in standard form.

15. $\dfrac{3-4i}{1+3i}$

15. _____

16. $-\dfrac{9}{3i}$

16. _____

Chapter 11 Quadratic Functions and Equations
11.1 Quadratic Functions and Their Graphs

Graphs of Quadratic Functions ~ Min-Max Applications ~ Basic Transformations of $y = ax^2$ ~
Transformations of $y = ax^2 + bx + c$ (Optional)

STUDY PLAN

Read: Read Section 11.1 on pages 714-724 in your textbook or eText.

Practice: Do your assigned exercises in your   ☐ Book   ☐ MyMathLab   ☐ Worksheets

Review: Keep your corrected assignments in an organized notebook and use them to review
for the test.

Key Terms
Exercises 1-6: Use the vocabulary terms listed below to complete each statement.
Note that some terms or expressions may not be used.

wider	**upward**
vertex	**narrower**
downward	**reflection**
vertical shift	**horizontal shift**
axis of symmetry	**quadratic function**

1. The _____ is the lowest point on the graph of a parabola that opens upward and
    the highest point on the graph of a parabola that opens downward.

2. In general, the graph of $y = -ax^2$ is a(n) _____ of the graph of $y = ax^2$ across
    the x-axis.

3. A(n) _____ is a vertical line about which a parabola is symmetric.

4. In the graph of $y = ax^2 + bx + c$, the value of c affects the placement of the parabola in the
    xy-plane. This placement is often called _____.

5. A(n) _____ can be written in the form $f(x) = ax^2 + bx + c$, where a, b, and c
    are constants with $a \neq 0$.

6. The graph of $y = ax^2$ is a parabola with the following characteristics.
    (a) The _____ is $(0,0)$ and the _____ is given by $x = 0$.
    (b) It opens _____ if $a > 0$ and opens _____ if $a < 0$.
    (c) It is _____ than the graph of $y = x^2$, if $0 < |a| < 1$.
    (d) It is _____ than the graph of $y = x^2$, if $|a| > 1$.

Graphs of Quadratic Functions

Exercises 1-6: Refer to Examples 1-3 on pages 715-717 in your text and the Section 11.1 lecture video.

Use the graph of the quadratic function to identify the vertex, the axis of symmetry, and whether the parabola opens upward or downward.

1.

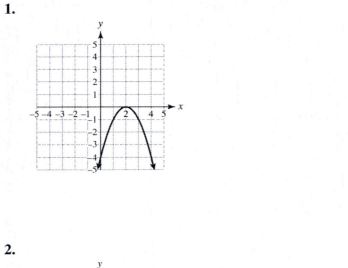

1. _____

2.

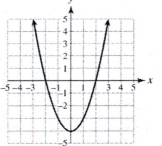

2. _____

3. Find the vertex for the graph of $f(x) = 3x^2 + 6x + 5$.
Support your answer graphically.

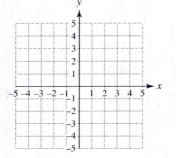

3. _____

Identify the vertex and axis of symmetry on the graph of $y = f(x)$. *Graph* $y = f(x)$.

4. $f(x) = x^2 - 2$

4. _____

5. $f(x) = -(x+4)^2$

5. _____

6. $f(x) = -x^2 - 2x$

6. _____

Min-Max Applications

Exercises 7-10: Refer to Examples 4-7 on pages 718-720 in your text and the Section 11.1 lecture video.

7. A printing company gives discounts for bulk sales of shirts. The regular price of a shirt is \$30, and the price decreases by \$0.50 for each shirt sold.

(a) Write a formula $f(x)$ that gives the total price for x shirts.

7.(a)_____

(b) Determine how many shirts should be sold to maximize the revenue.

(b)_____

8. Find the maximum y-value on the graph of $f(x) = -x^2 - 3x - 2$.

8. _____

9. A baseball is hit into the air and its height h in feet after t seconds can be calculated by $h(t) = -16t^2 + 96t + 4$.

(a) What is the height of the baseball when it is hit?

9.(a)_____

(b) Determine the maximum height of the baseball.

(b)_____

10. A hotel is considering giving the following group discount on room rates. The regular price for a room is $112, but for each room rented the price decreases by $4. A graph of the revenue received from renting x rooms is shown.

Hotel Revenue

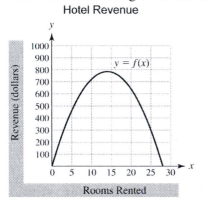

(a) Interpret the graph.

10.(a)_____

(b) What is the maximum revenue? How many rooms should be rented to receive the maximum revenue?

(b)_____

(c) Write a formula for $f(x)$, representing revenue.

(c)_____

(d) Use $f(x)$ to determine symbolically the maximum revenue and the number of rooms that should be rented.

(d)_____

Basic Transformations of $y = ax^2$

Exercise 11: Refer to Example 8 on page 721 in your text and the Section 11.1 lecture video.

11. Compare the graph of $g(x) = -\dfrac{1}{2}x^2$ to the graph of $f(x) = x^2$.

11. _____

Then graph both functions on the same coordinate axes.

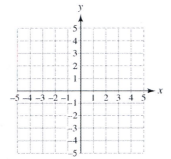

Transformations of $y = ax^2 + bx + c$ **(Optional)**

Exercise 12: Refer to Example 9 on pages 722-723 in your text and the Section 11.1 lecture video.

12. Let $f(x) = \dfrac{1}{2}x^2 - 2x - \dfrac{5}{2}$.

    (a) Does the graph of f open upward or downward?   12.(a)_____

        Is this graph wider or narrower than the graph of $y = x^2$?   _____

    (b) Find the axis of symmetry and the vertex.   **(b)**_____

        _____

    (c) Find the y-intercept and any x-intercepts.   **(c)**_____

        _____

    (d) Sketch a graph of f.

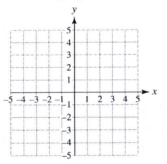

Chapter 11 Quadratic Functions and Equations
11.2 Parabolas and Modeling

Vertical and Horizontal Translations ~ Vertex Form ~ Modeling with Quadratic Functions (Optional)

STUDY PLAN

Read: Read Section 11.2 on pages 728-738 in your textbook or eText.

Practice: Do your assigned exercises in your ☐ Book ☐ MyMathLab ☐ Worksheets

Review: Keep your corrected assignments in an organized notebook and use them to review for the test.

Key Terms
Exercises 1-7: Use the vocabulary terms listed below to complete each statement.
Note that some terms or expressions may not be used.

vertex	$y = a(x-h)^2 + k$	$y = x^2 + k$
upward		
translations	$y = x^2 - k$	$y = (x+h)^2$
downward		
completing the square	$y = (x-h)^2$	

1. Let h and k be positive numbers. To graph _____, shift the graph of $y = x^2$ by h units right.

2. Shifts that change the position of a graph, but not its shape, are called _____.

3. Let h and k be positive numbers. To graph _____, shift the graph of $y = x^2$ by k units upward.

4. The _____ form of the equation of a parabola with vertex (h,k) is _____, where $a \neq 0$ is a constant. If $a > 0$, the parabola opens _____; if $a < 0$, the parabola opens _____.

5. Let h and k be positive numbers. To graph _____, shift the graph of $y = x^2$ by h units left.

6. The vertex of a parabola can be found by using a technique called _____.

7. Let h and k be positive numbers. To graph _____, shift the graph of $y = x^2$ by k units downward.

Vertical and Horizontal Translations

Exercises 1-3: Refer to Example 1 on page 730 in your text and the Section 11.2 lecture video.

Sketch the graph of the equation and identify the vertex.

1. $f(x) = x^2 + 3$ 1. _____

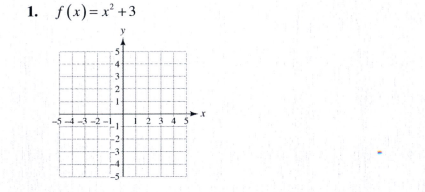

2. $f(x) = (x-3)^2$ 2. _____

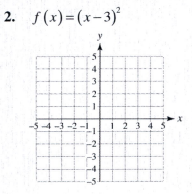

3. $f(x) = (x+2)^2 - 4$ 3. _____

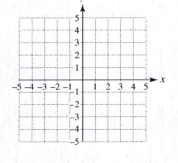

Vertex Form

Exercises 4-11: Refer to Examples 2-6 on pages 731-734 in your text and the Section 11.2 lecture video.

Compare the graph of $y = f(x)$ to the graph of $y = x^2$. Then sketch a graph of $y = f(x)$ and $y = x^2$ in the same xy-plane.

4. $f(x) = 2(x-1)^2 - 3$ 4. _____

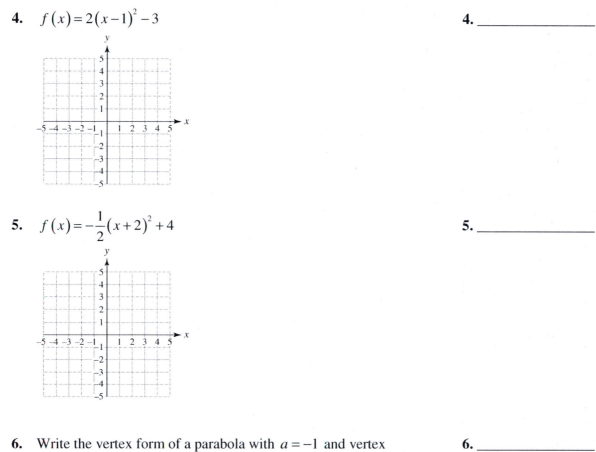

5. $f(x) = -\dfrac{1}{2}(x+2)^2 + 4$ 5. _____

6.   Write the vertex form of a parabola with $a = -1$ and vertex   6. _____
     $(-3,-1)$. Then express this equation in the form $y = ax^2 + bx + c$.   _____

7.   Find the vertex on the graph of $y = 2x^2 - 4x - 1$. Write this   7. _____
     equation in vertex form.   _____

8.   Write the equation $y = x^2 - 6x$ in vertex form by completing   8. _____
     the square. Identify the vertex.   _____

Write each equation in vertex form. Identify the vertex.

9. $y = x^2 + 4x + 5$ 9. _____

                                                                    _____

10. $y = x^2 + 5x - 2$ 10. _____

                                                                    _____

11. $y = -2x^2 + 6x - 4$ 11. _____

                                                                    _____

Modeling with Quadratic Functions (Optional)

Exercise 12: Refer to Example 7 on page 735 in your text and the Section 11.2 lecture video.

12.   Find a value for the constant a so that $f(x) = ax^2$ models the   12. _____
      data in the table.

x	-6	-3	0	3	6
$f(x)$	-12	-3	0	-3	-12

Chapter 11 Quadratic Functions and Equations
11.3 Quadratic Equations

Basics of Quadratic Equations ~ The Square Root Property ~ Completing the Square ~ Solving an Equation for a Variable ~ Applications of Quadratic Equations

---

STUDY PLAN

Read: Read Section 11.3 on pages 742-751 in your textbook or eText.

Practice: Do your assigned exercises in your ☐ Book ☐ MyMathLab ☐ Worksheets

Review: Keep your corrected assignments in an organized notebook and use them to review for the test.

Key Terms
Exercises 1-2: Use the vocabulary terms listed below to complete each statement.
Note that some terms or expressions may not be used.

   quadratic equation
   square root property

1. A(n) _____ can be written as $ax^2 + bx + c = 0,$ where a, b, and c are constants with $a \neq 0$.

2. The _____ states that if k is a nonnegative number, then the solutions to the equation $x^2 = k$ are given by $x = \pm\sqrt{k}$.

Basics of Quadratic Equations

Exercises 1-3: Refer to Example 1 on pages 743-745 in your text and the Section 11.3 lecture video.

Solve each quadratic equation. Support your results numerically and graphically.

1. $2x^2 + 5 = 0$ 1. _____

2. $x^2 + x - 6 = 0$ 2. _____

3. $x^2 + 1 = 2x$ 3. _____

The Square Root Property

Exercises 4-7: Refer to Examples 2-3 on page 746 in your text and the Section 11.3 lecture video.

Solve each equation.

4. $x^2 = 8$ 4. _____

5. $4x^2 - 25 = 0$ 5. _____

6. $(x+3)^2 = 16$ 6. _____

7. An object falls from a height of 80 feet. How long does it take   7. _____
    for the object to hit the ground?

Completing the Square

Exercises 8-10: Refer to Examples 4-6 on pages 747-748 in your text and the Section 11.3 lecture video.

8. Find the term that should be added to $x^2 + 8x$ to form a perfect square trinomial.

   8. _____

9. Solve the equation $x^2 + 4x - 3 = 0$.

   9. _____

10. Solve the equation $2x^2 - 5x = 4$.

    10. _____

Solving an Equation for a Variable

Exercises 11-12: Refer to Example 7 on page 749 in your text and the Section 11.3 lecture video.

Solve each equation for the specified variable.

11. $x = 9y^2 + 1$, for y

    11. _____

12. $V = \pi r^2 h$, for r
    $\left(\text{Hint: } r > 0.\right)$

    12. _____

Applications of Quadratic Equations

Exercises 13-14: Refer to Examples 8-9 on pages 749-750 in your text and the Section 11.3 lecture video.

13. Find a safe speed limit x for a curve with a radius of 200 feet by using the equation $R = \dfrac{1}{2}x^2$.

    13. _____

14. The function $f(x) = 0.0066x^2 - 23.76x + 21,389$ models the population of the United States in millions from 1800 through 2000, where $x = 1800$ corresponds to the year 1800, etc. Determine the approximate population of the United States in the year 1950.

    14. _____

Understanding Concepts through Multiple Approaches
(For additional practice, visit MyMathLab.)

15. Solve the equation $x^2 - 5x + 6 = 0$.

(a) Solve algebraically.

(b) Solve numerically using the table feature of a graphing calculator.

(c) Solve visually using a graphing calculator.

Did you get the same result using each method? Which method do you prefer? Explain why.

Chapter 11 Quadratic Functions and Equations
11.4 The Quadratic Formula

Solving Quadratic Equations ~ The Discriminant ~ Quadratic Equations Having Complex Solutions

---

STUDY PLAN

Read: Read Section 11.4 on pages 755-763 in your textbook or eText.

Practice: Do your assigned exercises in your ☐ Book ☐ MyMathLab ☐ Worksheets

Review: Keep your corrected assignments in an organized notebook and use them to review for the test.

Key Terms
Exercises 1-5: Use the vocabulary terms listed below to complete each statement.
Note that some terms or expressions may not be used.

    no
    one
    two
    discriminant
    quadratic formula

1. For the quadratic equation $ax^2 + bx + c = 0$, if $b^2 - 4ac = 0$, there is/are _____ real solution(s).

2. The solutions to $ax^2 + bx + c = 0$ with $a \neq 0$ are given by the _____,
$$x = \frac{-b \pm \sqrt{b^2 - 4ac}}{2a}.$$

3. For the quadratic equation $ax^2 + bx + c = 0$, if $b^2 - 4ac < 0$, there is/are _____ real solution(s).

4. The expression $b^2 - 4ac$ is called the _____.

5. For the quadratic equation $ax^2 + bx + c = 0$, if $b^2 - 4ac > 0$, there is/are _____ real solution(s).

Solving Quadratic Equations

Exercises 1-4: Refer to Examples 1-4 on pages 756-759 in your text and the Section 11.4 lecture video.

1. Solve the equation $4x^2 - 7x + 1 = 0$. Support your results graphically.

    1. _____

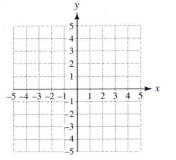

2. Solve the equation $9x^2 + 24x + 16 = 0$. Support your results graphically.

    2. _____

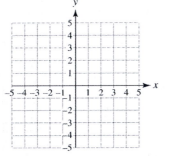

3. Solve the equation $5x^2 + 2x + 1 = 0$. Support your results graphically.

    3. _____

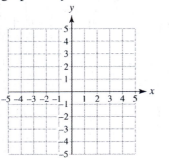

4. If a car's headlights do not illuminate the road beyond 600 feet, estimate a safe nighttime speed limit x for the car by solving $\frac{1}{9}x^2 + \frac{11}{3}x = 600$.

    4. _____

The Discriminant

Exercises 5-6: Refer to Examples 5-6 on page 760 in your text and the Section 11.4 lecture video.

5. A graph of $f(x) = ax^2 + bx + c$ is shown.

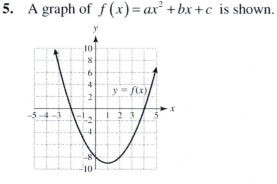

$y = f(x)$

   (a) State whether $a > 0$ or $a < 0$.   5.(a)_____

   (b) Solve the equation $ax^2 + bx + c = 0$.   (b)_____

   (c) Determine whether the discriminant is positive, negative, or zero.   (c)_____

6. Use the discriminant to determine the number of solutions to   6. _____
   $25x^2 + 9 = -30x$. Then solve the equation, using the quadratic
   formula.   _____

Quadratic Equations Having Complex Solutions

Exercises 7-10: Refer to Examples 7-10 on pages 761-762 in your text and the Section 11.4 lecture video.

7. Solve $x^2 + 7 = 0$.   7. _____

8. Solve $2x^2 + x + 4 = 0$. Write your answer in standard form: $a + bi$.   **8.** _____

9. Solve $\frac{1}{3}x^2 + 2 = x$. Write your answer in standard form: $a + bi$.   **9.** _____

10. Solve $x(x - 6) = -25$ by completing the square.   **10.** _____

Chapter 11 Quadratic Functions and Equations
11.5 Quadratic Inequalities

Basic Concepts ~ Graphical and Numerical Solutions ~ Symbolic Solutions

---

STUDY PLAN

Read: Read Section 11.5 on pages 768-774 in your textbook or eText.

Practice: Do your assigned exercises in your ☐ Book ☐ MyMathLab ☐ Worksheets

Review: Keep your corrected assignments in an organized notebook and use them to review for the test.

Key Terms
Exercises 1-3: Use the vocabulary terms or expressions listed below to complete each statement. Note that some terms or expressions may not be used.

$p < x < q$
$x < p$ **or** $x > p$
test value
quadratic inequality

1. If the equals sign in a quadratic equation is replaced with $>$, $\geq$, $<$. or $\leq$, a(n) _____ results.

2. A(n) _____ is substituted into an inequality to see if it results in a true statement.

3. Let $ax^2 + bx + c = 0$, $a > 0$, have two real solutions p and q, where $p < q$.
$ax^2 + bx + c < 0$ is equivalent to _____.
$ax^2 + bx + c > 0$ is equivalent to _____.
Quadratic inequalities involving $\leq$ or $\geq$ can be solved similarly.

Basic Concepts

Exercises 1-2: Refer to Example 1 on page 768 in your text and the Section 11.5 lecture video.

Determine whether the inequality is quadratic.

1. $4x + x^2 \geq -3$ 1. _____

2. $x^2(x+2) \geq 5$ 2. _____

Graphical and Numerical Solutions

Exercises 3-8: Refer to Examples 2-5 on pages 769-772 in your text and the Section 11.5 lecture video.

3. Make a table of values for $y = x^2 + x - 6$ and then sketch the graph.   3. _____

   Use the table and graph to solve $x^2 + x - 6 \leq 0$. Write your answer in interval notation.

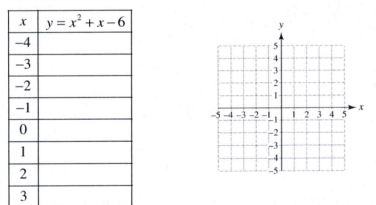

x	$y = x^2 + x - 6$
-4	
-3	
-2	
-1	
0	
1	
2	
3	

4. Solve $x^2 \geq 16$. Write your answer in interval notation.   4. _____

Solve each of the inequalities graphically.

5. $x^2 + 2 < 0$

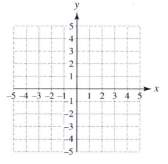

5. _____

6. $x^2 + 2 \geq 0$

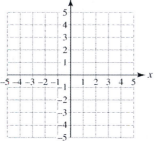

6. _____

7. $(x-2)^2 > 0$

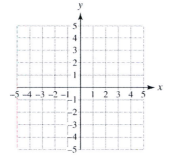

7. _____

8.   Suppose that the elevation E in feet of a sag, or sag curve, is given by $E(x) = 0.00004x^2 - 0.4x + 2000$, where x is the horizontal distance in feet along the sag curve and $0 \leq x \leq 10{,}000$. Estimate graphically the x-values where the elevation is 1200 feet or less.

8. _____

Symbolic Solutions

Exercises 9-11: Refer to Examples 6-7 on pages 773-774 in your text and the Section 11.5 lecture video.

Solve each inequality symbolically. Write your answer in interval notation.

9. $10 \le x(7-x)$ 9. _____

10. $6x^2 > x+1$ 10. _____

11. A rectangular corral is 10 feet longer than it is wide. The area of   11. _____
    the corral must be at least 600 square feet. What widths are possible?

Chapter 11 Quadratic Functions and Equations
11.6 Equations in Quadratic Form

Higher Degree Polynomial Equations ~ Equations Having Rational Exponents ~ Equations Having Complex Solutions

---

STUDY PLAN

> **Read:** Read Section 11.6 on pages 777-781 in your textbook or eText.
>
> **Practice:** Do your assigned exercises in your ☐ Book ☐ MyMathLab ☐ Worksheets
>
> **Review:** Keep your corrected assignments in an organized notebook and use them to review for the test.

Higher Degree Polynomial Equations

Exercise 1: Refer to Example 1 on page 778 in your text and the Section 11.6 lecture video.

1. Solve $4x^6 - x^3 = 3$.   1. _____

Equations Having Rational Exponents

Exercises 2-3: Refer to Examples 2-3 on pages 778-779 in your text and the Section 11.6 lecture video.

2. Solve $6x^{-2} - x^{-1} - 12 = 0$.   2. _____

3. Solve $2x^{2/3} + 5x^{1/3} + 2 = 0$.   3. _____

Equations Having Complex Solutions

Exercises 4-5: Refer to Examples 4-5 on page 779-780 in your text and the Section 11.6 lecture video.

4. Find all complex solutions to $x^4 - 81 = 0$.   **4.** _____

5. Find all complex solutions to $\dfrac{1}{x} + \dfrac{1}{x^2} = -2$.   **5.** _____

Chapter 12 Exponential and Logarithmic Functions
12.1 Composite and Inverse Functions

Composition of Functions ~ One-to-one Functions ~ Inverse Functions ~ Tables and Graphs of Inverse Functions

---

STUDY PLAN

Read: Read Section 12.1 on pages 795-805 in your textbook or eText.

Practice: Do your assigned exercises in your ☐ Book ☐ MyMathLab ☐ Worksheets

Review: Keep your corrected assignments in an organized notebook and use them to review for the test.

Key Terms
Exercises 1-5: Use the vocabulary terms listed below to complete each statement. Note that some terms or expressions may not be used. Some terms may be used more than once.

one-to-one
composition
vertical line test

inverse function
composite function
horizontal line test

1. The _____ is used to identify functions, whereas the _____ is used to identify one-to-one functions.

2. A function f is _____ if, for any c and d in the domain of f, $c \neq d$ implies that $f(c) \neq f(d)$. That is, different inputs always result in different outputs.

3. The _____ states that if every horizontal line interests the graph of a function f at most once, then f is a one-to-one function.

4. If f and g are functions, then the _____ $g \circ f$, or _____ of g and f, is defined by $(g \circ f)(x) = g(f(x))$.

5. Let f be a one-to-one function. Then f^{-1} is the _____ of f if $(f^{-1} \circ f)(x) = f^{-1}(f(x)) = x$, for every x in the domain of f, and $(f \circ f^{-1})(x) = f(f^{-1}(x)) = x$, for every x in the domain of f^{-1}.

Composition of Functions

Exercises 1-7: Refer to Examples 1-3 on pages 796-797 in your text and the Section 12.1 lecture video.

Evaluate $(g \circ f)(2)$ and then find a formula for $(g \circ f)(x)$.

1. $f(x) = x^3$; $g(x) = 2x - 6$ 1. _____

2. $f(x) = -3x$; $g(x) = x^2 + 2x - 5$ 2. _____

3. $f(x) = \sqrt{2x}$; $g(x) = \dfrac{1}{x-2}$ 3. _____

Use the tables to evaluate each expression.

x	0	1	2	3
$f(x)$	2	1	3	0

x	0	1	2	3
$g(x)$	0	3	1	2

4. $(f \circ g)(1)$ 4. _____

5. $(g \circ f)(2)$ 5. _____

6. $(f \circ f)(0)$ 6. _____

7.   Use the graph to evaluate $(g \circ f)(2)$.   7. _____

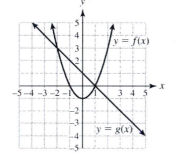

One-to-One Functions

Exercises 8-9: Refer to Example 4 on page 799 in your text and the Section 12.1 lecture video.

Determine whether each graph represents a one-to-one function.

8.   8. _____

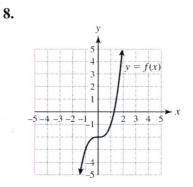

9.   9. _____

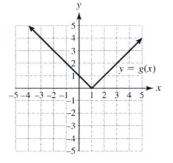

Inverse Functions

Exercises 10-14: Refer to Examples 5-7 on pages 800-802 in your text and the Section 12.1 lecture video.

Give the inverse operations for each statement. Then write a function f for the given statement and a function g for its inverse operation.

10.   Multiply x by -2.

10. _____

_____

_____

11.   Double x and then subtract 15 from the result.

11. _____

_____

_____

12.   Verify that $f^{-1}(x) = 2x - 1$ if $f(x) = \dfrac{x+1}{2}$.

12. _____

Find the inverse of each one-to-one function.

13. $f(x) = 4x - 1$

13. _____

14. $f(x) = (x-1)^3$

14. _____

Tables and Graphs of Inverse Functions

Exercises 15-16: Refer to Examples 8-9 on pages 803-804 in your text and the Section 12.1 lecture video.

15. The graph of $y = f(x)$ is shown. Sketch a graph of $y = f^{-1}(x)$.   **15.** _____

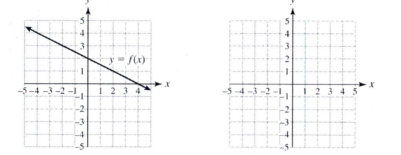

16. The line graph shown represents a function f.

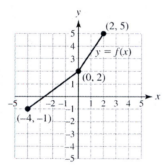

(a) Is f a one-to-one function?   **16.(a)** _____

(b) Sketch the graph of $y = f^{-1}(x)$.

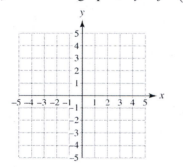

Chapter 12 Exponential and Logarithmic Functions
12.2 Exponential Functions

Basic Concepts ~ Graphs of Exponential Functions ~ Percent Change and Exponential Functions ~ Compound Interest ~ Models Involving Exponential Functions ~ The Natural Exponential Function

---

STUDY PLAN

Read: Read Section 12.2 on pages 809-821 in your textbook or eText.

Practice: Do your assigned exercises in your ☐ Book ☐ MyMathLab ☐ Worksheets

Review: Keep your corrected assignments in an organized notebook and use them to review for the test.

Key Terms
Exercises 1-6: Use the vocabulary terms listed below to complete each statement.
Note that some terms or expressions may not be used.

base   decay factor
exponential   coefficient
growth factor   compound interest
percent change   continuous growth
natural exponential

1. The equation $A = P(1 + r)^t$ represents _____. The _____ is $(1 + r)$.

2. The function represented by $f(x) = e^x$ is the _____ function, where $e \approx 2.71828$.

3. When an amount A changes to a new amount B, then the _____ is calculated by $\dfrac{B - A}{A} \times 100$.

4. A function represented by $f(x) = Ca^x$, with $a > 0$ and $a \neq 1$, is a(n) _____ function with _____ a and _____ C.

5. The natural exponential function is frequently used to model _____.

6. In the formula $f(x) = Ca^x$, a is called the _____ when $a > 1$ and the _____ when $0 < a < 1$.

Basic Concepts

Exercises 1-3: Refer to Example 1 on page 811 in your text and the Section 12.2 lecture video.

Evaluate $f(x)$ for the given value of x.

1. $f(x) = 4(5)^x$; $x = 2$

    1. _____

2. $f(x) = 4\left(\dfrac{1}{3}\right)^x$; $x = -2$

    2. _____

3. $f(x) = \dfrac{1}{2}(4)^x$; $x = -1$

    3. _____

Graph of Exponential Functions

Exercises 4-7: Refer to Examples 2-3 on pages 812-813 in your text and the Section 12.2 lecture video.

4. Compare $f(x) = 2^x$ and $g(x) = 2x$ graphically and numerically.

    (a) Graph $f(x) = 2^x$ and $g(x) = 2x$ on the same axes.

    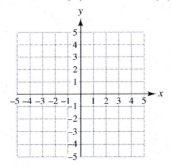

    (b) Complete the table.

x	−2	−1	0	1	2	3
$f(x)$						
$g(x)$						

For each table, determine whether f is a linear function or an exponential function.
Find a formula for f.

5.

x	−2	−1	0	1	2
$f(x)$	25	5	1	$\dfrac{1}{5}$	$\dfrac{1}{25}$

5._____

_____

6.

x	−2	−1	0	1	2
$f(x)$	−5	−3	−1	1	3

6._____

_____

7.

x	0	1	2	3	4
$f(x)$	$\dfrac{1}{4}$	$\dfrac{1}{2}$	1	2	4

7._____

_____

Percent Change and Exponential Functions

Exercises 8-10: Refer to Examples 4-6 on pages 814-816 in your text and the Section 12.2 lecture video.

8. Complete the following.

 (a) Find the percent change if an account balance increases   8.(a)_____
 from $800 to $1000.

 (b) Find the percent change if an account balance decreases   (b)_____
 from $1250 to $750.

9. An account that contains $5000 increases in value by 25%.

 (a) Find the increase in value of the account.   9.(a)_____

 (b) Find the final value of the account.   (b)_____

 (c) By what factor a did the account increase?   (c)_____

10. Initially a laboratory culture contains 80,000 live bacteria per
 milliliter and it is decreasing in numbers by 10% per hour.

 (a) Write the formula for an exponential function B that gives   10.(a)_____
 the number of bacteria per milliliter after x hours.

 (b) Evaluate $B(4)$ and interpret your results.   (b)_____

Compound Interest

Exercises 11-12: Refer to Examples 7-8 on pages 817-818 in your text and the Section 12.2 lecture video.

11. A 30-year old worker deposits $5000 in a retirement account that that pays 9% annual interest at the end of each year. How much money would be in the account when the worker is 60 years old? What is the growth factor?

11. _____

   _____

12. Initially, $2500 is deposited in an account paying 3% annual interest, compounded monthly. What is the account balance after 8 years?

12. _____

Models Involving Exponential Functions

Exercises 13-14: Refer to Examples 9-10 on pages 818-819 in your text and the Section 12.2 lecture video.

13. On average, a particular intersection has 280 vehicles arriving each hour. Highway engineers use the formula $f(x) = (0.955)^x$ to estimate the likelihood, or probability, that no vehicle will enter the intersection within an interval of x seconds.

   (a) Evaluate $f(10)$ and interpret the results.

   13. (a)_____

   (b) Is this function an example of exponential growth or exponential decay?

   (b)_____

14. In a particular forest, the probability or likelihood that no tree is located within a circle of radius x feet can be estimated by $P(x) = (0.785)^x$.

   (a) Evaluate $P(10)$ and interpret the result.

   14. (a)_____

   (b) What happens to $P(x)$ as x becomes large?

   (b)_____

The Natural Exponential Function

Exercise 15: Refer to Example 11 on page 820 in your text and the Section 12.2 lecture video.

15. In the year 2010 in a certain town, the population was 400,000 and was growing at a continuous rate of 2.5% per year. The population (in thousands) x years after 2010 can be modeled by $f(x) = 400e^{0.025x}$. Estimate the population in 2025 to the nearest thousand.

15. _____

Chapter 12 Exponential and Logarithmic Functions
12.3 Logarithmic Functions

The Common Logarithmic Function ~ The Inverse of the Common Logarithmic Function ~ Logarithms with Other Bases

---

STUDY PLAN

Read: Read Section 12.3 on pages 826-834 in your textbook or eText.

Practice: Do your assigned exercises in your ☐ Book ☐ MyMathLab ☐ Worksheets

Review: Keep your corrected assignments in an organized notebook and use them to review for the test.

Key Terms
Exercises 1-5: Use the vocabulary terms listed below to complete each statement.
Note that some terms or expressions may not be used.

natural logarithm   common logarithmic function
common logarithm   logarithmic function with base a
logarithm with base a

1. The _____ of a positive number x, denoted $\log x$, is calculated as follows. If x is written as $x = 10^k$, then $\log x = k$, where k is a real number. That is, $\log 10^k = k$.

2. The function given by $f(x) = \log_a x$ is called the _____.

3. A base-e logarithm is referred to as a(n) _____.

4. The _____ of a positive number x, denoted $\log_a x$, is calculated as follows. If x is written as $x = a^k$, then $\log_a x = k$, where $a > 0$, $a \neq 1$, and k is a real number. That is, $\log_a a^k = k$.

5. The function given by $f(x) = \log x$ is called the _____.

The Common Logarithmic Function

Exercises 1-6: Refer to Examples 1-2 on page 827 in your text and the Section 12.3 lecture video.

Evaluate each expression, if possible.

1. $\log 10,000$

1. _____

2. $\log(-5)$

2. _____

Evaluate each common logarithm.

3. $\log 1000$

3. _____

4. $\log \dfrac{1}{100}$

4. _____

5. $\log \sqrt{100}$

5. _____

6. $\log 85$

6. _____

The Inverse of the Common Logarithmic Function

Exercises 7-13: Refer to Examples 3-5 on pages 829-830 in your text and the Section 12.3 lecture video.

Use inverse properties to simplify each expression.

7. $\log 10^e$

7. _____

8. $\log 10^{x^2-1}$

8. _____

9. $10^{\log 3}$

9. _____

10. $10^{\log 8x}$, $x > 0$

10. _____

Graph each function f and compare its graph to $y = \log x$.

11. $f(x) = \log(x+3)$

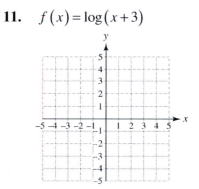

12. $f(x) = \log(x) - 2$

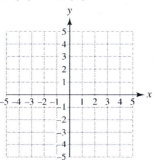

13. Sound levels in decibels (dB) can be computed by
$f(x) = 160 + 10\log x,$ where x is the intensity of the sound
in watts per square centimeter. The noise at a concert had an
intensity level of 10^{-3} w/cm^2. What decibel level is this?

13. _____

Logarithms with Other Bases

Exercises 14-26: Refer to Examples 6-10 on pages 831-834 in your text and the Section 12.3
lecture video.

Simplify each logarithm.

14. $\log_2 \dfrac{1}{8}$

14. _____

15. $\log_3 9$

15. _____

Approximate to the nearest hundredth.

16. $\ln 100$

16. _____

17. $\ln \dfrac{1}{5}$

17. _____

Simplify each logarithm.

18. $\log_3 27$ **18.** _____

19. $\log_5 \dfrac{1}{125}$ **19.** _____

20. $\log_8 1$ **20.** _____

21. $\log_2 16^{-1}$ **21.** _____

Simplify each expression.

22. $\ln e^{12x}$ **22.** _____

23. $e^{\ln 1.5}$ **23.** _____

24. $3^{\log_3 2x}$ **24.** _____

25. $10^{\log(3x+1)}$ **25.** _____

26. There is a mathematical relationship between an airplane's weight x and the runway length required at takeoff. For certain types of airplanes, the minimum runway length L in thousands of feet may be modeled by $L(x) = 1.3\ln x$, where x is in thousands of pounds.

(a) Estimate the runway length needed for an airplane weighing 15,000 pounds.

26.(a)_____

(b) Does a 30,000-pound airplane need twice the runway length that a 15,000-pound airplane needs? Explain.

(b)_____

Chapter 12 Exponential and Logarithmic Functions
12.4 Properties of Logarithms

Basic Properties ~ Change of Base Formula

STUDY PLAN

Read: Read Section 12.4 on pages 838-843 in your textbook or eText.

Practice: Do your assigned exercises in your ☐ Book ☐ MyMathLab ☐ Worksheets

Review: Keep your corrected assignments in an organized notebook and use them to review for the test.

Key Terms
Exercises 1-4: Use the expressions listed below to complete each statement.
Note that some expressions may not be used.

$\log_a(rm)$ $\log_a m \cdot \log_a n$ $\dfrac{\ln x}{\ln a}$

$r \log_a m$ $\log_a m + \log_a n$

$\log_a m - \log_a n$ $\dfrac{\log x}{\log a}$ $\dfrac{\log_a m}{\log_a n}$

1. For positive numbers m, n, and $a \neq 1$, $\log_a mn = $ _____.

2. For positive numbers m, n, and $a \neq 1$, $\log_a \dfrac{m}{n} = $ _____.

3. For positive numbers m and $a \neq 1$ and any real number r, $\log_a(m^r) = $ _____.

4. Let x and $a \neq 1$ be positive real numbers. Then $\log_a x = $ _____ or $\log_a x = $ _____.

Basic Properties

Exercises 1-22: Refer to Examples 1-8 on pages 839-842 in your text and the Section 12.4 lecture video.

Write each expression as a sum of logarithms. Assume that x is positive.

1. $\log 33$ 1. _____

2. $\ln 7x$ 2. _____

3. $\ln x^3$ 3. _____

Write each expression as one logarithm. Assume that x and y are positive.

4. $\log_2 3 + \log_2 4$ 4. _____

5. $\log xy + \log y$ 5. _____

6. $\ln 4x + \ln 2x$ 6. _____

Write each expression as a difference of two logarithms. Assume that variables are positive.

7. $\ln\dfrac{5}{2}$

7. _____

8. $\log\dfrac{x}{4y}$

8. _____

9. $\log_2\dfrac{y^3}{z}$

9. _____

Write each expression as one term. Assume that x is positive.

10. $\ln 16 - \ln 4$

10. _____

11. $\log_5 y^4 - \log_5 y$

11. _____

12. $\log 24x - \log 4x$

12. _____

Rewrite each expression, using the power rule.

13. $\ln 3^4$

13. _____

14. $\log (1.4)^{x+2}$

14. _____

15. $\log_3 7^{2x}$

15. _____

Write each expression as the logarithm of a single expression.

16. $2\ln x + \ln x^4$

16. _____

17. $3\log x - \log \sqrt{x}$

17. _____

Write each expression in terms of logarithms of x, y and z.

18. $\ln \dfrac{x\sqrt[3]{y}}{z^2}$

18. _____

19. $\log \sqrt{\dfrac{y}{xz}}$

19. _____

Using only properties of logarithms and the approximations $\ln 2 \approx 0.7,$ $\ln 3 \approx 1.1,$ *and* $\ln 5 \approx 1.6,$ *find an approximation for each expression.*

20. $\ln 6$ **20.** _____

21. $\log 25$ **21.** _____

22. $\ln \dfrac{6}{5}$ **22.** _____

Change of Base Formula

Exercise 23: Refer to Example 9 on page 842 in your text and the Section 12.4 lecture video.

23. Approximate $\log_3 12$ to the nearest thousandth.   **23.** _____

Chapter 12 Exponential and Logarithmic Functions
12.5 Exponential and Logarithmic Equations

Exponential Equations and Models ~ Logarithmic Equations and Models

---

STUDY PLAN

Read: Read Section 12.5 on pages 845-852 in your textbook or eText.

Practice: Do your assigned exercises in your   ☐ Book ☐ MyMathLab ☐ Worksheets

Review: Keep your corrected assignments in an organized notebook and use them to review for the test.

Exponential Equations and Models

Exercises 1-11: Refer to Examples 1-5 on pages 846-849 in your text and the Section 12.5 lecture video.

Solve and approximate to the nearest hundredth.

1. $10^x = 180$ 1. _____

2. $e^x = 50$ 2. _____

3. $2^x = 40$ 3. _____

4. $0.8^x = 0.5$ 4. _____

Solve each equation and approximate to the nearest hundredth.

5. $2e^x + 3 = 7$

5. _____

6. $4^{x-2} = 16$

6. _____

7. $e^{3x} = e^{x-4}$

7. _____

8. $3^{4x} = 2^{x+1}$

8. _____

9. Graphs for $f(x) = 0.4e^x$ and $g(x) = 3$ are shown.

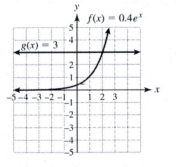

(a) Use the graphs to estimate the solution to the equation $f(x) = g(x)$.

9. (a)_____

(b) Check your estimate by solving the equation symbolically.

(b)_____

10. Solve $10(2.3)^x = 210$ symbolically. Round answer to the   10. _____
nearest hundredth.

11. The life spans of 129 robins were monitored over a 4-year period
in one study. The formula $f(x) = 10^{-0.42x}$ can be used to calculate
the percentage of robins remaining after x years.

(a) Evaluate $f(0.5)$ and interpret the result.   11. (a)_____

(b) Determine when 35% of the robins remained.   (b)_____

Logarithmic Equations and Models

Exercises 12-17: Refer to Examples 6-9 on pages 849-852 in your text and the Section 12.5 lecture video.

Solve and approximate to the nearest hundredth when appropriate.

12. $3\log x = 9$ 12. _____

13. $\ln 6x = 7.2$ 13. _____

14. $\log_2(x-3) = 5$ 14. _____

15. Solve $\log(x+3) + \log(x-3) = \log 7$. Check any answers.   15. _____

16. For some types of airplanes with weight x, the minimum   **16.** _____
runway length L required at takeoff is modeled by
$L(x) = 3\log x$. In this equation, L is measured in thousands of
feet and x is measured in thousand of pounds. Estimate
the weight of the heaviest airplane that can take off from
a runway that is 4200 feet long.

Understanding Concepts through Multiple Approaches
(For additional practice, visit MyMathLab.)

17. Two functions, f and g, are given.

$f(x) = 5\log x$

$g(x) = 3$

(a) Solve $f(x) = g(x)$ algebraically.

(b) Solve visually using the graph shown.

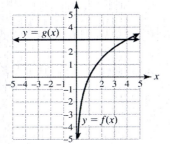

Did you get the same result using each method? Which method do you prefer? Explain why.

Chapter 13 Conic Sections
13.1 Parabolas and Circles

Types of Conic Sections ~ Parabolas with Horizontal Axes of Symmetry ~ Equations of Circles

---

STUDY PLAN

Read: Read Section 13.1 on pages 870-876 in your textbook or eText.

Practice: Do your assigned exercises in your ☐ Book ☐ MyMathLab ☐ Worksheets

Review: Keep your corrected assignments in an organized notebook and use them to review for the test.

Key Terms
Exercises 1-3: Use the vocabulary terms listed below to complete each statement.
Note that some terms or expressions may not be used.

> center
> circle
> radius
> conic section
> standard equation of a circle

1. _____(s) are named after the different ways that a plane can intersect a cone.

2. The _____ with center (h,k) and radius r is $(x-h)^2 + (y-k)^2 = r^2$.

3. A(n) _____ consists of the set of points in a plane that are the same distance from a fixed point. The fixed distance is called the _____, and the fixed point is called the _____.

Parabolas with Horizontal Axes of Symmetry

Exercises 1-4: Refer to Examples 1-3 on pages 872-873 in your text and the Section 13.1 lecture video.

1.   Graph $x = -3y^2$. Find its vertex and axis of symmetry.   1. _____

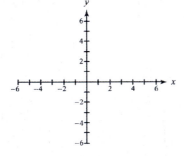

2.   Graph $x = (y+2)^2 - 3$. Find its vertex and axis of symmetry.   2. _____

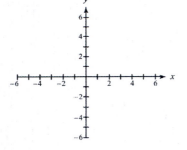

Identify the vertex and then graph each parabola.

3. $x = (y+3)^2$ 3. _____

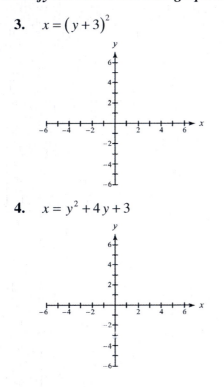

4. $x = y^2 + 4y + 3$ 4. _____

Equations of Circles

Exercises 5-7: Refer to Examples 4-6 on pages 874-875 in your text and the Section 13.1 lecture video.

5. Graph $x^2 + y^2 = 4$. Find the radius and center.   5. _____

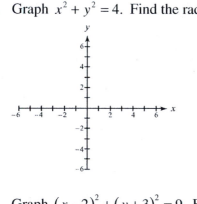

6. Graph $(x-2)^2 + (y+3)^2 = 9$. Find the radius and center.   6. _____

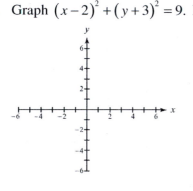

7. Find the center and radius of the circle given by   7. _____
    $x^2 - 2x + y^2 + 6y = 6$.

Chapter 13 Conic Sections
13.2 Ellipses and Hyperbolas

Equations of Ellipses ~ Equations of Hyperbolas

STUDY PLAN

Read: Read Section 13.2 on pages 879-886 in your textbook or eText.

Practice: Do your assigned exercises in your ☐ Book ☐ MyMathLab ☐ Worksheets

Review: Keep your corrected assignments in an organized notebook and use them to review for the test.

Key Terms
Exercises 1-7: Use the vocabulary terms listed below to complete each statement.
Note that some terms or expressions may not be used. Some terms may be used more than once.

vertices   focus
center   ellipse
branches   asymptotes
major axis   hyperbola
fundamental rectangle   transverse axis

1. A(n) _____ is the set of points in a plane the difference of whose distances from two fixed points is constant. Each fixed point is called a(n) _____.

2. The dashed line $y = \pm \dfrac{b}{a}x$ or $y = \pm \dfrac{a}{b}x$ is a(n) _____ for a hyperbola centered at the origin.

3. The _____ of an ellipse is the midpoint of the major axis.

4. The line segment connecting the vertices of a hyperbola is called the _____.

5. A(n) _____ is the set of points in a plane the sum of whose distances from two fixed points is constant. Each fixed point is called a(n) _____.

6. A hyperbola consists of two curves, or _____.

7. The _____ of an ellipse are located at the endpoints of the major axis.

Equations of Ellipses

Exercises 1-4: Refer to Examples 1-2 on pages 881-882 in your text and the Section 13.2 lecture video.

Sketch a graph of each ellipse. Label the vertices and endpoints of the minor axes.

1. $\dfrac{x^2}{4} + \dfrac{y^2}{9} = 1$

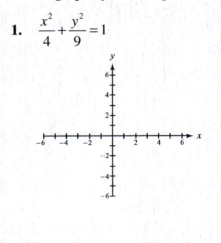

2. $25x^2 + 9y^2 = 225$

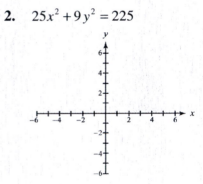

Use the graph to determine the standard equation of the ellipse.

3.

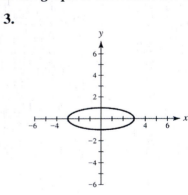

3. _____

4.

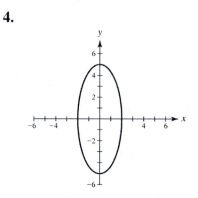

4. _____

Equations of Hyperbolas

Exercises 5-6: Refer to Examples 4-5 on pages 885-886 in your text and the Section 13.2 lecture video.

5. Sketch a graph of $\dfrac{x^2}{4} - \dfrac{y^2}{16} = 1$. Label the vertices and show the asymptotes.

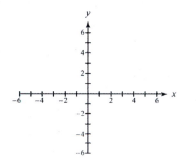

6. Use the graph to determine the standard equation of the hyperbola.   6. _____

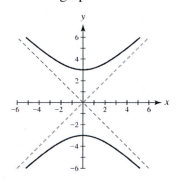

Chapter 13 Conic Sections
13.3 Nonlinear Systems of Equations and Inequalities

Basic Concepts ~ Solving Nonlinear Systems of Equations ~ Solving Nonlinear Systems of Inequalities

---

STUDY PLAN

Read: Read Section 13.3 on pages 890-896 in your textbook or eText.

Practice: Do your assigned exercises in your ☐ Book ☐ MyMathLab ☐ Worksheets

Review: Keep your corrected assignments in an organized notebook and use them to review for the test.

Key Terms
Exercises 1-3: Use the vocabulary terms listed below to complete each statement.
Note that some terms or expressions may not be used. Some terms may be used more than once.

   method of substitution
   nonlinear system of equations
   nonlinear system of inequalities

1. The system $x^2 + y^2 = 5$ is a(n) _____.
                $y = 2x$

2. A(n) _____ can be solved using graphical techniques. The solution consists of all points in the intersection of the shaded regions.

3. The _____ is a symbolic technique used to solve nonlinear systems of equations.

Solving Nonlinear Systems of Equations

Exercises 1-4: Refer to Examples 1-4 on pages 891-893 in your text and the Section 13.3 lecture video.

Solve the nonlinear system of equations symbolically and graphically.

1. $y = 3x$

   $x^2 + y^2 = 40$

   1. _____

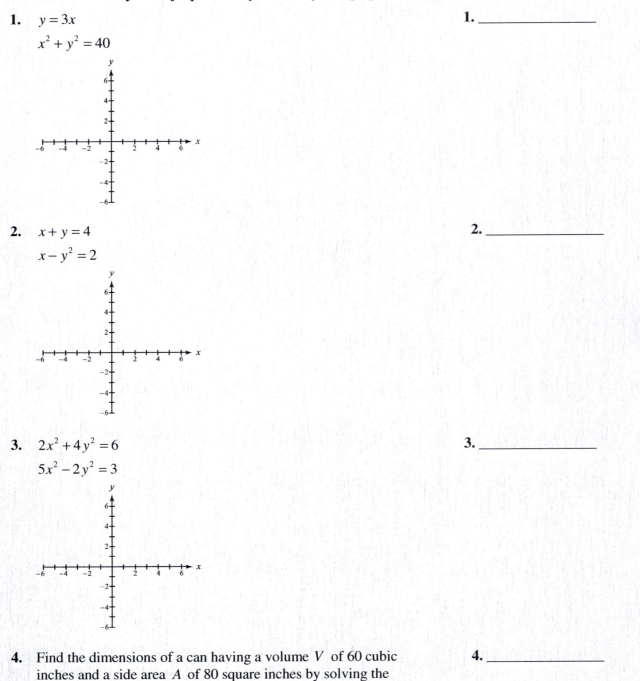

2. $x + y = 4$

   $x - y^2 = 2$

   2. _____

3. $2x^2 + 4y^2 = 6$

   $5x^2 - 2y^2 = 3$

   3. _____

4. Find the dimensions of a can having a volume V of 60 cubic inches and a side area A of 80 square inches by solving the following nonlinear system of equations symbolically.

   4. _____

   _____

   $\pi r^2 h = 60$

   $2\pi rh = 80$

Solving Nonlinear Systems of Inequalities

Exercises 5-7: Refer to Examples 5-7 on pages 894-895 in your text and the Section 13.3 lecture video.

Shade the solution set for the system of inequalities.

5. $\dfrac{x^2}{9} + \dfrac{y^2}{16} \leq 1$

$y < 1$

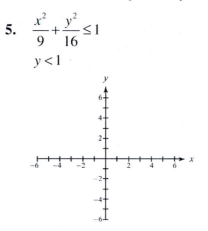

6. $\dfrac{y^2}{4} - \dfrac{x^2}{9} \geq 1$

$\dfrac{x^2}{16} + \dfrac{y^2}{9} \leq 1$

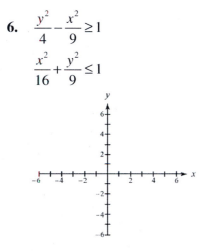

7. $2x \geq y^2$

$y > x - 2$

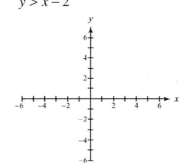

Understanding Concepts through Multiple Approaches
(For additional practice, visit MyMathLab.)

8.   Solve the system of equations.

$$x - y = -1$$
$$x^2 - 3y = -3$$

(a)   Solve algebraically.

(b)   Solve numerically.

x	0	1	2	3	4
$y =$					
$y =$					

(c)   Solve visually.

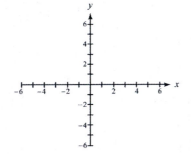

Did you get the same result using each method? Which method do you prefer? Explain why.

Chapter 14 Sequences and Series
14.1 Sequences

Basic Concepts ~ Representations of Sequences ~ Models and Applications

STUDY PLAN

  Read: Read Section 14.1 on pages 909-915 in your textbook or eText.

  Practice: Do your assigned exercises in your ☐ Book ☐ MyMathLab ☐ Worksheets

  Review: Keep your corrected assignments in an organized notebook and use them to review for the test.

Key Terms
Exercises 1-3: Use the vocabulary terms listed below to complete each statement.
Note that some terms or expressions may not be used.

   nth term
   finite sequence
   general term
   infinite sequence

1. A(n) _____ is a function whose domain is the set of natural numbers.

2. The _____, or _____, of a sequence is $a_n = f(n)$.

3. A(n) _____ is a function whose domain is $D = \{1,2,3,...,n\}$ for some fixed natural number n.

Basic Concepts

Exercises 1-3: Refer to Example 1 on page 911 in your text and the Section 14.1 lecture video.

Write the first four terms of each sequence for n = 1, 2, 3, and 4.

1. $f(n) = \dfrac{1}{n+2}$

1. _____

2. $f(n) = 4(-1)^n$

2. _____

3. $f(n) = 5n - 1$

3. _____

Representations of Sequences

Exercises 4-5: Refer to Examples 2-3 on page 912 in your text and the Section 14.1 lecture video.

4.   Use the figure to write the terms of the sequence.

4. _____

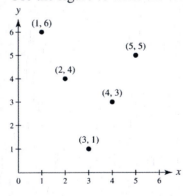

5. A driver travels at an average speed of 65 mph. Give symbolic, numerical, and graphical representations for a sequence that models the distance driven over a 5-hour period.

5. _____

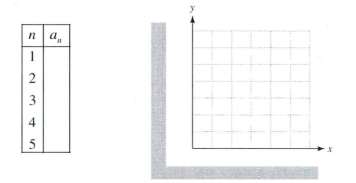

n	a_n
1	
2	
3	
4	
5	

Models and Applications

Exercise 6: Refer to Example 4 on page 913 in your text and the Section 14.1 lecture video.

6. Suppose the initial population of a small town is 30,000 and that the population is growing at a rate of 10% per year.

    (a) Find the general term of a sequence that represents the population in a given year.

    6. (a)_____

    (b) Represent numerically the population over a 5-year period.

    (b)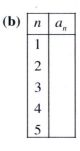

n	a_n
1	
2	
3	
4	
5	

Chapter 14 Sequences and Series
14.2 Arithmetic and Geometric Sequences

Representations of Arithmetic Sequences ~ Representations of Geometric Sequences ~ Applications and Models

---

STUDY PLAN

Read: Read Section 14.2 on pages 917-923 in your textbook or eText.

Practice: Do your assigned exercises in your ☐ Book ☐ MyMathLab ☐ Worksheets

Review: Keep your corrected assignments in an organized notebook and use them to review for the test.

Key Terms
Exercises 1-3: Use the vocabulary terms listed below to complete each statement.
Note that some terms or expressions may not be used. Some terms may be used more than once.

common ratio
arithmetic sequence
common difference
geometric sequence

1. A(n) _____ is a linear function given by $a_n = dn + c$ whose domain is the set of natural numbers. The value of d is called the _____.

2. A(n) _____ is given by $a_n = a_1 (r)^{n-1}$, where n is a natural number and $r \neq 0$ or 1. The value of r is called the _____, and a_1 is the first term of the sequence.

3. The nth term of a(n) _____ is given by $a_n = a_1 + (n-1)d$, where a_1 is the first term and d is the _____.

Representations of Arithmetic Sequences

Exercises 1-6: Refer to Examples 1-3 on pages 918-919 in your text and the Section 14.2 lecture video.

Determine whether f is an arithmetic sequence. If it is, identify the common difference d.

1. $f(n) = 3 - 4n$ 1. _____

2.
n	$f(n)$
1	−4
2	−1
3	0
4	1
5	4

2. _____

3. A graph of f is shown.   3. _____

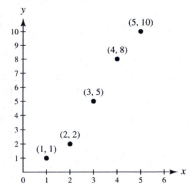

Find the general term a_n for each arithmetic sequence.

4. $a_1 = -1$ and $d = 5$ 4. _____

5. $a_1 = 1$ and $a_4 = -8$ 5. _____

6. If $a_1 = 4$ and $d = 2$, find a_{48}.   6. _____

Representations of Geometric Sequences

Exercises 7-12: Refer to Examples 4-6 on pages 920-921 in your text and the Section 14.2 lecture video.

Determine whether f is a geometric sequence. If it is, identify the common ratio.

7. $f(n) = -3(0.05)^{n-1}$

7. _____

8.

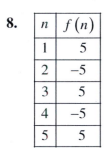

n	$f(n)$
1	5
2	−5
3	5
4	−5
5	5

8. _____

9. A graph of f is shown.

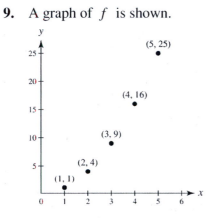

9. _____

Find the general term a_n for each geometric sequence.

10. $a_1 = -11$ and $r = \dfrac{1}{2}$

10. _____

11. $a_1 = 64$ and $a_4 = 1$

11. _____

12. If $a_1 = 3$ and $r = -2$, find a_{12}.

12. _____

Applications and Models

Exercises 13-14: Refer to Examples 7-8 on pages 921-922 in your text and the Section 14.2 lecture video.

13. Suppose that an employee has an initial salary of $58,000 and receives a $3000 raise each year.

    (a) Write a sequence that gives the employee's annual salary for years 1, 2, 3, 4 and 5. Is this sequence arithmetic, geometric, or neither?   13.(a)_____

    (b) Write the general term for this sequence.   (b)_____

    (c) Find a_{10} and interpret the result.   (c)_____

14. Suppose that the water in a swimming pool initially has a chlorine content of 4 parts per million and that 25% of the chlorine dissipates each day.

    (a) Write a sequence that models the amount of chlorine in the pool at the beginning of the first 3 days, assuming that no additional chorine is added. Is this sequence arithmetic, geometric, or neither?   14.(a)_____

    (b) Write the general term for this sequence.   (b)_____

    (c) At the beginning of what day does the chorine first drop below 1 part per million?   (c)_____

Chapter 14 Sequences and Series
14.3 Series

Basic Concepts ~ Arithmetic Series ~ Geometric Series ~ Summation Notation

---

STUDY PLAN

Read: Read Section 14.3 on pages 926-933 in your textbook or eText.

Practice: Do your assigned exercises in your ☐ Book ☐ MyMathLab ☐ Worksheets

Review: Keep your corrected assignments in an organized notebook and use them to review for the test.

Key Terms
Exercises 1-6: Use the vocabulary terms listed below to complete each statement.
Note that some terms or expressions may not be used.

annuity   upper limit
lower limit   arithmetic sequence
arithmetic series   geometric sequence
summation notation   index of summation
finite series

1. A(n) _____ is an expression of the form $a_1 + a_2 + a_3 + \ldots + a_n$.

2. A sum of money from which regular payments are made is called a(n) _____.

3. The sum of the first n terms of a(n) _____, denoted S_n, is found by averaging the first and nth terms and then multiplying by n. That is, $S_n = a_1 + a_2 + a_3 + \ldots + a_n = n\left(\dfrac{a_1 + a_n}{2}\right)$.

4. The expression $\displaystyle\sum_{k=1}^{n} a_k = a_1 + a_2 + a_3 + \ldots + a_n$ is referred to as _____.
   The letter k is called the _____.
   The number 1 is called the _____.
   The number n is called the _____.

5. Summing the terms of an arithmetic sequence results in a(n) _____.

6. If its first term is a_1 and its common ratio is r, then the sum of the first n terms of a(n) _____ is given by $S_n = a_1\left(\dfrac{1 - r^n}{1 - r}\right)$, provided $r \neq 1$.

Basic Concepts

Exercise 1: Refer to Example 1 on page 927 in your text and the Section 14.3 lecture video.

1. The table presents a sequence a_n that computes the number of new internet users in a particular area from 2001 to 2007, where $n = 1$ corresponds to the year 2001.

n	1	2	3	4	5	6	7
a_n	27,133	26,980	24,219	24,069	25,196	20,110	22,337

(a) Write a series whose sum represents the total number of new internet users from 2001 to 2007. Find the sum.

1. (a)_____

(b) Interpret $a_1 + a_2 + a_3 + \ldots + a_{10}$.

(b)_____

Arithmetic Series

Exercises 2-3: Refer to Examples 2-3 on pages 928-929 in your text and the Section 14.3 lecture video.

2. Suppose that a person has a starting annual salary of $45,000 and receives a $2000 raise each year. Calculate the total amount earned after 12 years.

2. _____

3. Find the sum of the series $3 + 6 + 9 + \ldots + 45$.

3. _____

Geometric Series

Exercises 4-6: Refer to Examples 4-5 on pages 930-931 in your text and the Section 14.3 lecture video.

Find the sum of each series.

4. $1 - 3 + 9 - 27 + 81 - 243 + 729$

4. _____

5. $4 + 2 + 1 + \dfrac{1}{2} + \dfrac{1}{4} + \dfrac{1}{8} + \dfrac{1}{16} + \dfrac{1}{32}$

5. _____

6. Suppose that a 30-year old worker deposits $2000 into an annuity account at the end of each year. If the interest rate is 9%, find the future value of the annuity when the worker is 60 years old.

6. _____

Summation Notation

Exercises 7-11: Refer to Examples 6-8 on pages 931-932 in your text and the Section 14.3 lecture video.

Find each sum.

7. $\displaystyle\sum_{k=1}^{10} k^2$

7. _____

8. $\displaystyle\sum_{k=1}^{7} 4$

8. _____

9. $\displaystyle\sum_{k=4}^{10} (3k-1)$

9. _____

10. Express $3^3 + 4^3 + 5^3 + 6^3 + 7^3 + 8^3 + 9^3 + 10^3$ in summation notation.

10. _____

11. Suppose that a water filter removes 85% of the impurities entering it.

    (a) Find a series that represents the amount of impurities removed by a sequence of n water filters. Express this answer in summation notation.

11.(a)_____

    (b) How many water filters would be necessary to remove 97.75% of the impurities?

(b)_____

Chapter 14 Sequences and Series
14.4 The Binomial Theorem

Pascal's Triangle ~ Factorial Notation and Binomial Coefficients ~ Using the Binomial Theorem

---

STUDY PLAN

   Read: Read Section 14.4 on pages 936-940 in your textbook or eText.

   Practice: Do your assigned exercises in your   ☐ Book ☐ MyMathLab ☐ Worksheets

   Review: Keep your corrected assignments in an organized notebook and use them to review
           for the test.

Pascal's Triangle

Exercises 1-2: Refer to Example 1 on page 937 in your text and the Section 14.4 lecture video.

Expand each binomial, using Pascal's triangle.

   1. $(x-3)^5$ 1. _____

   2. $(2r+s)^3$ 2. _____

Factorial Notation and Binomial Coefficients

Exercises 3-5: Refer to Examples 2-3 on page 938 in your text and the Section 14.4 lecture video.

Simplify the expression.

   3. $\dfrac{5!}{1!0!}$ 3. _____

   4. $\dfrac{7!}{3!4!}$ 4. _____

5. Calculate ${}_4C_r$ for $r = 0, 1, 2, 3, 4$ by hand. Check your results on a calculator. Compare these numbers with the fourth row in Pascal's triangle.

5. _____

Using the Binomial Theorem

Exercises 6-8: Refer to Examples 4-5 on pages 939-940 in your text and the Section 14.4 lecture video.

Use the binomial theorem to expand each expression.

6. $(x - y)^6$

6. _____

7. $(2x - 5)^4$

7. _____

8. Find the third term of $(x + y)^6$.

8. _____

1.1 Numbers, Variables, and Expressions

Key Terms

1. formula

2. whole numbers

3. prime number

4. product; factors

5. prime factorization

6. variable

7. natural numbers

8. algebraic expression

9. composite number

10. equation

Prime Numbers and Composite Numbers

1. prime

2. neither

3. composite; $65 = 5 \cdot 13$

4. composite; $180 = 2 \cdot 2 \cdot 3 \cdot 3 \cdot 5$

Variables, Algebraic Expressions, and Equations

5. 10

6. 12

7. 15

8. 2

9. 20

10. -6

11. 3

12. 15

13. 1

14. 24

Translating Words to Expressions

15. $n - 5$; n is the number

16. $3d$; d is the cost of a DVD

17. $y(x - 5)$; x is one number, y is the other number

18. $\dfrac{100}{n}$; n is the number

19. (a) $1.50
    (b) $P = 0.15t$
    (c) $3.00

20. (a) $V = lwh$
    (b) 180 cubic inches

1.2 Fractions

Key Terms

1. lowest terms

2. least common denominator (LCD)

3. basic principle of fractions

4. multiplicative inverse; reciprocal

5. greatest common factor (GCF)

Basic Concepts

1. numerator: 7
   denominator: 15

2. numerator: x
   denominator: yz

3. numerator: $a + 2$
   denominator: $b - 3$

Simplifying Fractions to Lowest Terms

4. 14

5. 8

6. $\dfrac{2}{5}$

7. $\dfrac{3}{7}$

Multiplication and Division of Fractions

8. $\dfrac{12}{35}$

9. $\dfrac{4}{7}$

10. $\dfrac{5}{2}$

11. $\dfrac{ac}{5b}$

12. $\dfrac{1}{4}$

13. $\dfrac{2}{15}$

14. 6

15. $\dfrac{3}{10}$

16. $\dfrac{7}{16}$

17. 1

18. $\dfrac{5}{2}$

19. $\dfrac{xy}{3z}$

20. Answers will vary.

Addition and Subtraction of Fractions

21. $\dfrac{12}{11}$

22. $\dfrac{2}{3}$

23. 12

24. 60

25. $\dfrac{8}{12}, \dfrac{3}{12}$

26. $\dfrac{16}{60}, \dfrac{9}{60}$

27. $\dfrac{11}{12}$

28. $\dfrac{5}{12}$

29. $\dfrac{41}{42}$

An Application

30. $6\dfrac{7}{8}$ inches

1.3   Exponents and Order of Operations

Key Terms

1. exponential expression

2. base; exponent

3. parentheses, absolute values
exponential expressions
multiplication; division
addition; subtraction

Natural Number Exponents

1. 6^5

2. $\left(\dfrac{1}{3}\right)^4$

3. a^7

4. 64

5. 100,000

6. $\dfrac{8}{27}$

7. 10^3

8. 2^5

9. 3^4

Order of Operations

10. 3

11. 8

12. 7

13. $\dfrac{1}{5}$

14. 6

15. 11

16. 1

17. 28

Translating Words to Expressions

18. $3^3 - 8 = 19$

19. $30 + 4 \cdot 2 = 38$

20. $\dfrac{4^3}{2^2} = 16$

21. $\dfrac{40}{10-2} = 5$

1.4   Real Numbers and the Number Line

Key Terms

1. irrational number

2. principal square root

3. origin

4. average

5. rational number

6. opposite; additive inverse

7. square root

8. absolute value

9. integers

10. real number

Signed Numbers

1. -11

2. $\dfrac{3}{8}$

3. -5

4. $-\dfrac{1}{5}$

Integers and Rational Numbers

5. natural number,
whole number,
integer,
rational number

6. integer,
rational number

7. whole number,
integer,
rational number

8. rational number

Square Roots

9. 8

10. 20

11. 2.449

Real and Irrational Numbers

12. irrational number

13. integer,
rational number

14. rational number

15. natural number,
whole number,
integer,
rational number

16. Average height: 69 inches
natural number, rational number

The Number Line

17.

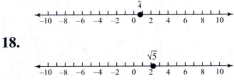

18.

19.

Absolute Value

20. 6.7

21. 4

22. 7

Inequality

23. $-3, -\sqrt{3}, 0, 2.2, \pi$

1.5   Addition and Subtraction of Real Numbers

Key Terms

1. less than or equal to; $a \leq b$

2. difference

3. greater than; $a > b$

4. approximately equal

5. greater than or equal to; $a \geq b$

6. addends; sum

7. less than; $a < b$

Addition of Real Numbers

1. $-32; 0$

2. $\sqrt{3}; 0$

3. $-\dfrac{5}{6}, 0$

4. -13

5. $\dfrac{1}{20}$

6. -4.3

7. 8

8. −3

9. −4

10. −10

Subtraction of Real Numbers

11. −20

12. −12

13. −2.4

14. $\dfrac{2}{15}$

15. 9

16. $\dfrac{1}{4}$

17. −6.4

Applications

18. 172°F

19. $265

1.6 Multiplication and Division of Real Numbers

Key Terms

1. negative

2. reciprocal; muliplicative inverse

3. positive

4. dividend; divisor; quotient

Multiplication of Real Numbers

1. −24

2. $\dfrac{3}{10}$

3. 4.8

4. −72

5. 25

6. −25

7. −64

8. −64

Division of Real Numbers

9. −45

10. $\dfrac{1}{9}$

11. $-\dfrac{3}{8}$

12. 0

13. $0.8\overline{3}$

14. 0.4375

15. 1.125

16. $\dfrac{2}{25}$

17. $\dfrac{11}{40}$

18. $\dfrac{1}{200}$

19. $1.\overline{6}$; $1\dfrac{2}{3}$

20. 0.9375; $\dfrac{15}{16}$

Applications

21. **(a)** $560,000

     (b) $320,000

22. 0.076

1.7 Properties of Real Numbers

Key Terms

1. identity property of 0

2. associative property for addition

3. multiplicative inverse property

4. distributive property

5. commutative property for multiplication

6. identity property of 1

7. associative property for multiplication

8. additive inverse property

9. commutative property for addition

Commutative Properties

1. $20 + 4$

2. $9 \cdot x$ or $9x$

Associative Properties

3. $2 + (4 + 7)$

4. $(ab)c$

5. associative property for addition

6. commutative property for multiplication

7. commutative property for addition

Distributive Properties

8. $4a - 20$

9. $-3b - 24$

10. $-x + 9$

11. $11 - y$

12. $(7 + 4)x = 11x$

13. $(2 - 10)a = -8a$

14. $(-6 + 3)y = -3y$

15. commutative and associative properties for addition

16. associative property for multiplication

17. distributive property

18. distributive property and commutative property for addition

Identity and Inverse Properties

19. identity property of 1

20. identity property of 0

21. multiplicative inverse property and identity property of 1

22. additive inverse property and identity property of 0

Mental Calculations

23. 1

24. 50

25. 195

26. 533

27. 5000 cubic feet

1.8   Simplifying and Writing Algebraic Expressions

Key Terms

1. additive identity

2. term

3. coefficient

4. like terms

5. multiplicative identity

Terms

1. Yes; -2

2. Yes; 4

3. No

4. Yes; 6

Combining Like Terms

5. unlike

6. like

7. unlike

8. like

9. $-\dfrac{7}{2}x$

10. $6y^2$

11. Cannot be combined

Simplifying Expressions

12. $-5 + 3x$

13. $7y - 13$

14. a

15. $-3b - 2$

16. $8x^2$

17. $-6t^3$

18. $4z - 3$

19. $5a - 5$

Writing Expressions

20. **(a)** $450w + 600w + 520w + 700w$
    $= 2270w$

    (b) $108,960 \text{ ft}^2$

Chapter 2 Linear Equations and Inequalities

2.1   Introduction to Equations

Key Terms

1. addition property of equality

2. solution set

3. multiplication property of equality

4. equivalent

5. solution

The Addition Property of Equality

1. -6

2. 11

3. $\dfrac{7}{6}$

4. 1

The Multiplication Property of Equality

5. -9

6. -5

7. 9

8. $-\dfrac{1}{2}$

9. (a) $G = 24x$
    (b) 25 years

2.2   Linear Equations

Key Terms

1. linear equation

2. no solutions

3. infinitely many solutions

4. contradiction

5. one solution

6. identity

Basic Concepts

1. Yes; $a = 3, b = -2$

2. Not linear

3. Not linear

4. Yes; $a = \dfrac{2}{3}, b = -6$

Solving Linear Equations

5. -2

x	-3	-2	-1	0	1	2	3
$-2x-5$	1	-1	-3	-5	-7	-9	-11

6. $\dfrac{7}{4}$

7. -6

8. $\dfrac{5}{2}$

9. The year 2010

Applying the Distributive Property

10. $-\dfrac{15}{4}$

11. $\dfrac{13}{2}$

Clearing Fractions and Decimals

12. $-\dfrac{14}{15}$

13. $\dfrac{3}{4}$

14. 2.24

15. 0.4375

Equations with No Solutions or Infinitely Many Solutions

16. Infinitely many solutions

17. One solution

18. No solutions

2.3   Introduction to Problem Solving

Key Terms

1. percent change

2. fraction; decimal number

3. add, plus, more, sum, total, increase

4. subtract, minus, less, difference, fewer, decrease

5. multiply, times, twice, double, triple, product

6. divide, divided by, quotient, per

7. equals, is, gives, results in, is the same as

Steps for Solving a Problem

1. $4x + 5 = 17;\ 3$

2. $\dfrac{1}{3}x + 7 = 1;\ -18$

3. $15 = 2x - 3;\ 9$

4. 17, 18, 19

5. 16,000

Percent Problems

6. $\dfrac{33}{100}$, 0.33

7. $\dfrac{1}{40}$, 0.025

8. $\dfrac{1}{125}$, 0.008

9. 13.7%

10. 12.5%

11. 161%

12. 20%

13. $55,775

14. 9.375 billion

Distance Problems

15. 570 mph

16. 8 mph

Other Types of Problems

17. 30 oz

18. $5000 at 9%; $1500 at 11%

2.4   Formulas

Key Terms

1. area

2. area

3. volume

4. circumference

5. perimeter

6. degree

7. area

8. circumference

Formulas from Geometry

1. (a) 123,500 square feet
   (b) approximately 2.8 acres

2. 36°; 72°; 72°

3. Circumference: $15\pi \approx 47.1$ centimeters
   Area: $56.25\pi \approx 177$ square centimeters

4. 10,275 square centimeters

5. Volume: 120 cubic meters
   Surface Area: 188 square meters

6. (a) approximately 19.63 cubic inches
   (b) approximately 10.9 fluid ounces

Solving for a Variable

7. (a) $l = \dfrac{P - 2w}{2}$
   (b) $l = 8$ inches

8. $h = \dfrac{2A}{a + b}$

9. $x = \dfrac{yz}{z - y}$

Other Formulas

10. 2.58

11. (a) $C = \dfrac{5}{9}(F - 32)$
    (b) 25°C

2.5   Linear Inequalities

Key Terms

1. interval notation

2. solution set

3. set-builder notation

4. linear inequality

5. solution

Solutions and Number Line Graphs

1.

2.

3.

4.

5. $(-\infty, 7)$

6. $[-4, \infty)$

7. $(0, \infty)$

8. Yes

9. No

10. $x = 0$

11. $x < 0$

12. $x \le 0$

13. $x > 0$

The Addition Property of Inequalities

14. $x \leq -9$

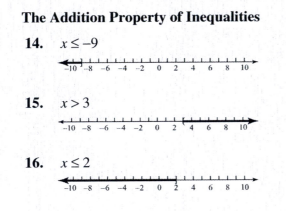

15. $x > 3$

16. $x \leq 2$

The Multiplication Property of Inequalities

17. $x \geq -7$

18. $x > 9$

19. $\left\{ x \middle| x < \dfrac{2}{3} \right\}$

20. $\left\{ x \middle| x \leq 8 \right\}$

21. $\left\{ x \middle| x > -3 \right\}$

Applications

22. $x < -10$

23. $x \leq 90$

24. $x \geq 3.5$

25. Altitudes greater than 4 miles

26. (a) $C = 4200 + 180x$
    (b) $R = 240x$
    (c) $P = 60x - 4200$
    (d) 70 items

Chapter 3 Graphing Equations

3.1 Introduction to Graphing

Key Terms

1. ordered pair

2. line graph

3. *x*-coordinate; *y*-coordinate

4. origin

5. scatterplot

6. quadrants

7. rectangular coordinate system

8. *x*-axis; *y*-axis

The Rectangular Coordinate System

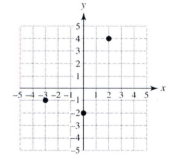

1. I

2. III

3. None; *y*-axis

4. 16 hours; 27 hours

5.

Price of a Gallon of Milk

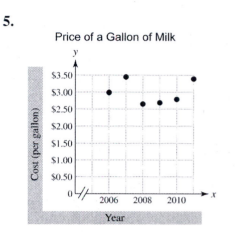

6.

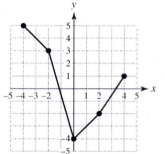

7. **(a)** no
   (a) 1960: about 115
   2000: about 150
   (a) approximately 30%

3.2 Linear Equations in Two Variables

Key Terms

1. linear equation in two variables

2. infinitely many

3. table

4. solution

5. standard form

Basic Concepts

1.   No

2.   No

3.   Yes

Tables of Solutions

4.
x	-4	-2	0	2
y	-8	-2	4	10

5.
x	3	0	-3	-6
y	0	4	8	12

6.   (a)
t	2	4	6	8	10
P	12.0	12.8	13.6	14.4	15.2

   (b)   14.4 million

Graphing Linear Equations in Two Variables

7.
x	-2	-1	0	1
y	6	3	0	-3

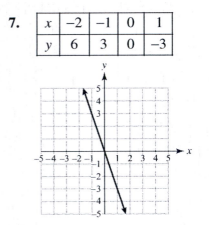

8. $y = -\dfrac{1}{2}x + 3$

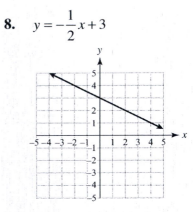

9. $x + y = -2$

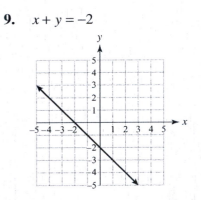

10. $y = \dfrac{3}{5}x - 3$

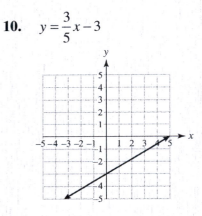

3.3   More Graphing of Lines

Key Terms

1.   horizontal line

2. y-coordinate; y-axis; $x = 0$; y

3.   vertical line

4. x-coordinate; x-axis; $y = 0$; x

Finding Intercepts

1. $-5x + 2y = 10$

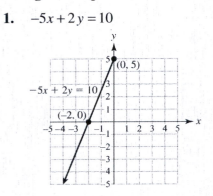

2.

x	-2	-1	0	1	2
y	0	1	2	3	4

x-intercept: -2
y-intercept: 2

3. (a)

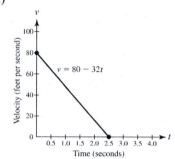

(b) The t-intercept indicates that the ball had a velocity of 0 feet per second after 2.5 seconds. The v-intercept indicates that the ball's initial velocity was 80 feet per second.

Horizontal Lines

4. y-intercept: -3

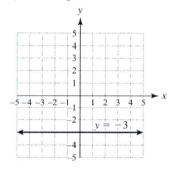

Vertical Lines

5. x-intercept: -1

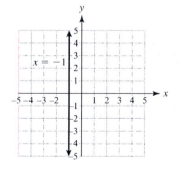

6. $y = 2$

7. $x = -3$

8. $x = -4$

9. $y = 1$

10. $x = 0$

3.4   Slope and Rates of Change

Key Terms

1.   zero slope

2.   negative slope

3.   run

4.   slope

5.   rate of change

6.   rise

7.   positive slope

8.   undefined slope

Finding Slopes of Lines

1. $m = -1$

2. $m = 0$

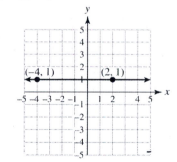

3. $m = 2$

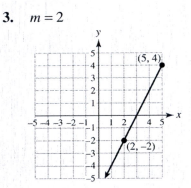

4. m is undefined

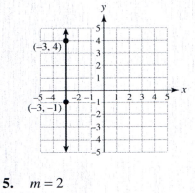

5. $m = 2$

6. m is undefined

7.

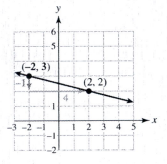

8.

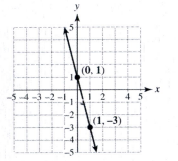

9.

x	-2	-1	0	1
y	4	1	-2	-5

Slope as a Rate of Change

10.   **(a)** y-intercept: 80
The cyclist is initially
80 miles from home.
(b) After 2 hours the cyclist
is 48 miles from home.
(c) $m = 16$; The cyclist is biking
at 16 miles per hour.

11.   **(a)** $m = 30$
(b) Profit increases, on average,
by \$30 for each additional
game consolefter made.

12.   **(a)**

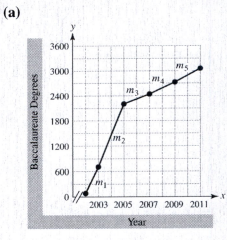

(b) $m_1 = 622$
$m_2 = 750$
$m_3 = 118.5$
$m_4 = 141.5$
$m_5 = 165.5$

(c) The slope indicates the increase
in the number of Baccalaureate
degrees awared.

13. (a)

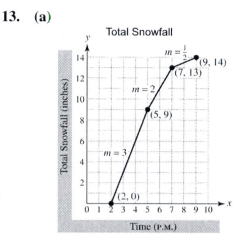

Total Snowfall

(b) The slope of each line segment represents the rate at which snow is falling.

3.5 Slope-Intercept Form

Key Terms

1. zero slope

2. parallel

3. slope

4. perpendicular

5. slope-intercept

6. negative reciprocals

7. undefined slope

Finding Slope-Intercept Form

1. $y = -2x + 2$

2. $y = x + 3$

3. $y = \dfrac{1}{3}x - 4$

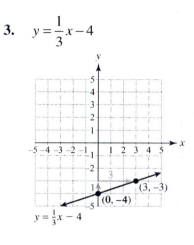

$$y = \tfrac{1}{3}x - 4$$

4. $y = \dfrac{5}{3}x + 5$; $m = \dfrac{5}{3}$; y-int: 5

5. $y = -\dfrac{1}{3}x + 2$; $m = -\dfrac{1}{3}$; y-int: 2

6. $y = 3x - 2$

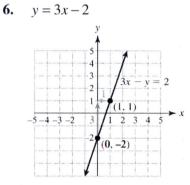

7. (a) \$274,000
   (b) $y = 120x + 34{,}000$
   (c) 3500

Parallel and Perpendicular Lines

8. $y = 2x + 5$

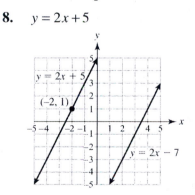

9. $y = \dfrac{1}{4}x$

10. $y = -\dfrac{3}{2}x$

11. $y = \dfrac{2}{5}x$

12. $y = -\dfrac{4}{3}x + 3$

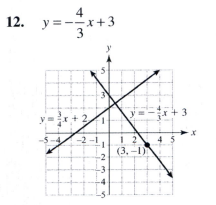

3.6 Point-Slope Form

Key Terms

1. point-slope

2. slope

3. slope-intercept

Derivation of Point-Slope Form

Writing Exercise: Answers will vary.

Finding Point-Slope Form

1. $y - 3 = -\dfrac{1}{2}(x + 2);$

$y = -\dfrac{1}{2}x + 2$

2. $y - 0 = 3(x - 0);$

$y = 3x$

3. $y + 2 = -\dfrac{1}{4}(x - 2);$ No

4. $y + 4 = -5(x - 2)$ or
$y - 1 = -5(x - 1)$

5. $y = -\dfrac{1}{2}x + 1$

6. $y = \dfrac{1}{4}x + 1$

7. $y = \dfrac{4}{3}x - 3$

8. $y = -3x + 1$

Applications

9. **(a)** $y - 56 = 3(x - 2006)$ or
$y - 74 = 3(x - 2010)$

(b) Slope $m = 3$ indicates that the salary increased, on average, by $3000 per year.

(c) $83,000

10. **(a)** 750 gallons per hour

(b) $y = -750x + 6000$

(c) The y-intercept is 6000 and indicates that the tank initially contained 6000 gallons of water. The x-intercept is 8 and indicates that the tank is empty after 8 hours.

(d)

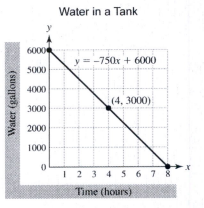

Water in a Tank

(e) After 4 hours, 3000 gallons of water remain in the tank.

3.7   Introduction to Modeling

Basic Concepts

1.   No, at $x = 6$, $D = 610$.

Modeling Linear Data

2.   **(a)**

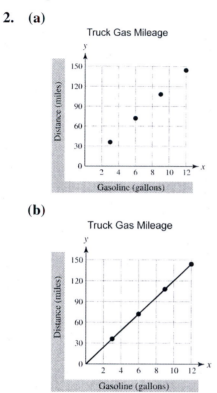

(b)

(c) $y = 12x$

The mileage of this truck is 12 miles per gallon.

(d)   192 miles

3.   **(a)** No, the points are not collinear.

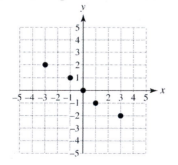

3.   **(b)** $y = -\dfrac{2}{3}x$

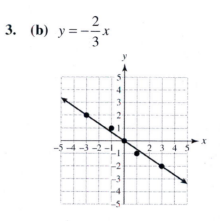

4.   **(a)** $N = 1.9x + 18.8$
        (b)   Aproximately 26.4 million

5. $y = 2500$

6. $y = 28x + 700$

7. $y = -10x + 480$

4.1 Solving Systems of Linear Equations Graphically and Numerically

Key Terms

1. consistent; independent; intersecting

2. intersection-of-graphs

3. inconsistent; parallel

4. solution to a system

5. system of linear equations

6. consistent; dependent; identical

Basic Concepts

1. 5 hours

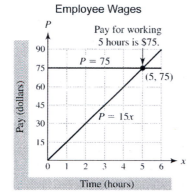

2. 2

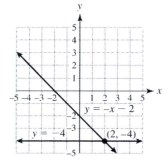

3. 0

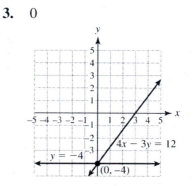

Solutions to Systems of Equations

4. infinitely many; consistent; dependent

5. 0; inconsistent

6. 1; consistent; independent

7. $(-5, -3)$

8. $(0, 3)$

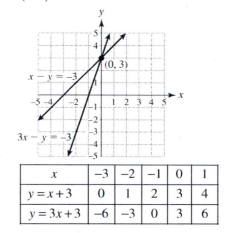

x	−3	−2	−1	0	1
$y = x + 3$	0	1	2	3	4
$y = 3x + 3$	−6	−3	0	3	6

9. $(1,-2)$

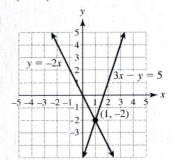

10.   5000 male students
7000 female students

4.2   Solving Systems of Linear Equations by Substitution

Key Terms

1.   method of substitution

2.   no solutions; parallel

3.   infinitely many solutions; identical

The Method of Substitution

1. $(-2,-4)$

2. $\left(2,\dfrac{1}{2}\right)$

3. $(5,0)$

4. $(2,1)$

Recognizing Other Types of Systems

5.   infinitely many solutions

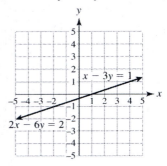

6.   no solutions

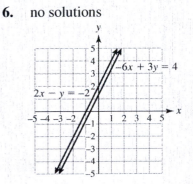

Applications

7.   300 radio ads
150 television ads

8.   hors d'oeuvres: $900
desserts: $1600

9.   speed of airplane: 450 mph
speed of wind: 30 mph

4.3   Solving Systems of Linear Equations by Elimination

Key Terms

1.   no solutions; inconsistent

2.   infinitely many solutions; consistent; dependent

3.   elimination method

The Elimination Method

1. $(-2,1)$

2. $(2,2)$

3. $(-4,3)$

4. $\left(-\dfrac{1}{4},\dfrac{3}{4}\right)$

5. $(3,-5)$

6. $(1, -3)$

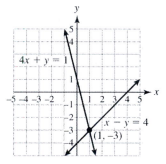

x	-2	-1	0	1	2
$y = x - 4$	-6	-5	-4	-3	-2
$y = -4x + 1$	9	5	1	-3	-7

Recognizing Other Types of Systems

7.   no solutions

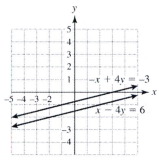

8.   infinitely many solutions

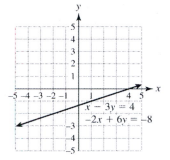

Applications

9.   (a) 1100 women
        (b) 700 men

10.   20 minutes on treadmill
          10 minutes on elliptical machine

Understanding Concepts through Multiple Approaches

11.   (a) $(-1, 4)$

         (b)

x	-2	-1	0	1	2
$y = -2x + 2$	6	4	2	0	-2
$y = 2x + 6$	2	4	6	8	10

         $(-1, 4)$

         (c) $(-1, 4)$

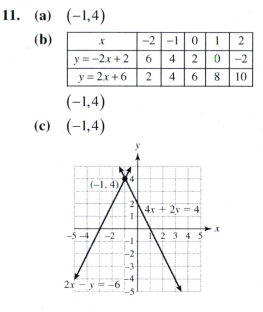

4.4   Systems of Linear Inequalities

Key Terms

1.   test point

2.   linear inequality

3.   solution

4.   system of linear inequalities

5.   solution set

Solutions to One Inequality

1. $y \leq 2$

2. $y > -2x$

3.

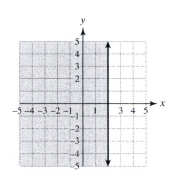

4.

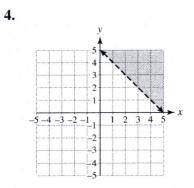

5.

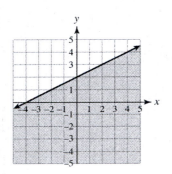

Solutions to Systems of Inequalities

6.

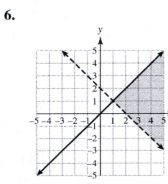

7.

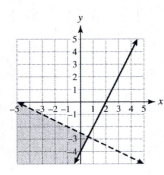

Applications

8.

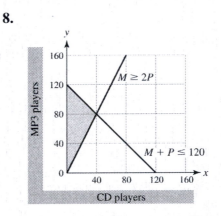

9. **(a)** The weight falls slightly outside of the recommended guidelines.
   (b) 150 to 200 lb

Chapter 5 Polynomials and Exponents

5.1 Rules for Exponents

Key Terms

1. power to a power

2. quotient to a power

3. product

4. base; exponent

5. product to a power

6. zero exponent; undefined

Review of Bases and Exponents

1. 7

2. $\dfrac{1}{4}$

3. -81

4. 16

Zero Exponents

5. -1

6. 4

7. 1

The Product Rule

8. 243

9. x^8

10. $12x^7$

11. $3x^6 + 2x^5$

12. a^5

13. $(x+y)^4$

Power Rules

14. 2^{15}

15. x^{16}

16. $125t^3$

17. $-8b^9$

18. $3x^{12}y^8$

19. $-64z^{18}$

20. $\dfrac{27}{64}$

21. $\dfrac{x^6}{y^6}$

22. $\dfrac{36}{(a-b)^2}$

23. $200x^5$

24. $\dfrac{a^9 b^3}{c^6}$

25. $-72x^{14}y^{12}$

26. (a) 3^2
    (b) $945,000$

5.2   Addition and Subtraction of Polynomials

Key Terms

1. trinomial

2. degree

3. polynomial

4. like

5. monomial

6. polynomial

7. unlike

8. coefficient

9. binomial

10. degree

Monomials and Polynomials

1. Yes; 3; 1; 3

2. Yes; 4; 2; 5

3. No

Addition of Polynomials

4. Unlike

5. Like; $4x^3y$

6. Like; $\dfrac{7}{2}xy^2$

7. $-x+6$

8. $5y^2+3y+8$

9. $12x^3-10x^2+2x-1$

10. $4b^2-4b+13$

Subtraction of Polynomials

11. $3x-2$

12. $-x^2-10x+1$

13. $4x^3+5x^2-6x-4$

14. $3x^2+x-6$

Evaluating Polynomial Expressions

15. $4xy$; 320 in^3

16. 37 cents

5.3   Multiplication of Polynomials

Key Terms

1. coefficients
   variables
   product

2. product; like terms

3. binomials

Multiplying Monomials

1. $-12x^5$

2. $-20a^4b^6$

Review of the Distributive Properties

3. $6x-3$

4. $8x^2-10$

5. $-5x^2+2x$

Multiplying Monomials and Polynomials

6. $15x^3-20x$

7. $4x^3-x^2$

8. $-6x^2+8x-4$

9. $3x^6 - 6x^5 + 9x^3$

10. $20x^4 y^3 + 4xy$

11. $-x^3 y - xy^3$

Multiplying Polynomials

12. $x^2 + 8x + 15$

13. $2x^2 - 5x + 3$

14. $3 + 11x - 4x^2$

15. $3x^3 + x^2 - 2x$

16. $4x^3 - x^2 + x + 3$

17. $4a^3 + 4a^2 b - 5ab^2 - 5b^3$

18. $b^6 - 4b^4 + 4b^2 - 1$

19. $x^3 + x^2 - 2x - 8$

20. (a) $h(h-2)(h+5) = h^3 + 3h^2 - 10h$
      (b) 3900 cubic inches

5.4 Special Products
Key Terms

1. $a^2 - b^2$

2. $a^2 + 2ab + b^2$

3. $a^2 - 2ab + b^2$

Product of a Sum and Difference

1. $a^2 - 1$

2. $z^2 - 9$

3. $9r^2 - 25t^2$

4. $9m^4 - 4n^4$

5. 391

Squaring Binomials

6. $x^2 - 8x + 16$

7. $9x^2 - 12x + 14$

8. $1 + 12b + 36b^2$

9. $9a^4 - 6a^2 b + b^2$

10. (a) $x^2 + 24x + 144$
      (b) The area of the pool and sidewalk
          is 1024 ft^2.

Cubing Binomials

11. $125x^3 + 150x^2 + 60x + 8$

12. (a) $1 + 3r + 3r^2 + r^3$
      (b) 1.125; The sum of money will
          increase by a factor of 1.125.

5.5 Integer Exponents and
the Quotient Rule

Key Terms

1. $\dfrac{1}{a^n}$; reciprocal

2. a^{m-n}

3. (a) a^n

    (b) $\dfrac{b^m}{a^n}$

    (c) $\left(\dfrac{b}{a}\right)^n$

4. scientific notation

Negative Integers as Exponents

1. $\dfrac{1}{16}$

2. $\dfrac{1}{5}$

3. $\dfrac{1}{x^3}$

4. $\dfrac{1}{(a+b)^4}$

5. $\dfrac{1}{27}$

6. $\dfrac{1}{625}$

7. $\dfrac{1}{x^2}$

8. $\dfrac{1}{t^6}$

9. $\dfrac{1}{a^4b^4}$

10. $\dfrac{1}{x^3y^{11}}$

The Quotient Rule

11. $\dfrac{1}{9}$

12. $\dfrac{3}{a^4}$

13. $\dfrac{x}{y^5}$

Other Rules for Exponents

14. 81

15. $\dfrac{27}{16}$

16. $\dfrac{x^5}{3y^7}$

17. $\dfrac{25}{b^6}$

Scientific Notation

18. 417,000

19. 0.00023

20. 0.0582

21. 3.2×10^7

22. 2×10^{-5}

23. About 1.2×10^7 round trips

5.6   Division of Polynomials

Division by a Monomial

1. $x^4 - x$

2. $\dfrac{b^5}{2} - \dfrac{3b^2}{2}$

3. $3x - 1 + \dfrac{2}{x}$

4. $3a + 4 - \dfrac{2}{a}$

5. $l = x + 3$

Division by a Polynomial

6. $3x - 2 + \dfrac{-2}{2x+3}$

7. $4x^2 + 4x + 2 + \dfrac{7}{x-1}$

8. $x - 4 + \dfrac{x+4}{x^2+4}$

Chapter 6 Factoring Polynomials and Solving Equations

6.1   Introduction to Factoring
Key Terms

1. factor

2. greatest common factor (GCF)

3. completely factored

Common Factors

1. $6(5x+3)$

2. $3x(3x-4)$

3. $4x(3x+2)$

4. $2y^2(5y-1)$

5. $4a(a^2-3a-2)$

6. $2x^2y(3x-y)$

7. $2a;\ 2a(2a+3)$

8. $5x^2;\ 5x^2(x^2-3)$

9. $3a^2b^2;\ 3a^2b^2(2a-5b)$

10. $8t(5-2t)$

Factoring by Grouping

11. $(x-3)(4x-5)$

12. $(2t+3)(t^2+6)$

13. $(3x^2+4)(x-2)$

14. $(4+a)(x-y)$

15. $(5x^2-1)(x-3)$

16. $(2z^3-3)(z+6)$

17. $4(x^2+1)(2x-3)$

18. $x^2(2x^2-5)(2x+3)$

6.2   Factoring Trinomials I
$$\left(x^2+bx+c\right)$$

Key Terms

1. standard form; leading coefficient

2. prime polynomial

Factoring Trinomials with Leading Coefficient 1

1. 4, 7

2. –8, 5

3. $(x+2)(x+5)$

4. $(x+3)(x+6)$

5. $(y+6)(y+7)$

6. $(b-3)(b-7)$

7. $(x-2)(x-6)$

8. $(y-5)(y+4)$

9. $(t+5)(t-8)$

10. $(x+6)(x-4)$

11. $(x-4)(x-3)$

12. Prime

13. $(x-7)(x+2)$

14. $5(x+2)(x+4)$

15. $2x^2(x+6)(x-1)$

16. Length: $x+5$;
Width: $x-2$

6.3   Factoring Trinomials II
(ax^2+bx+c)

Factoring Trinomials by Grouping

1. $(2x+3)(x+5)$

2. $(4a+1)(a-3)$

3. $(5x-2)(2x-5)$

4. Prime

5. Prime

6. $5(y-4)(3y+1)$

7. $3x(x^2-6x-9)$

Factoring with FOIL in Reverse

8. $(3a+4)(a-5)$

9. $(2x+1)(x+5)$

10. $(x+8)(5x+1)$

11. $-(3x-1)(2x+5)$

12. $-(2x+7)(x-5)$

6.4   Special Types of Factoring

Key Terms

1. $(a+b)(a^2-ab+b^2)$

2. $(a-b)^2$

3. $(a-b)(a^2+ab+b^2)$

4. $(a+b)^2$

5. $(a-b)(a+b)$

6. $a^2+2ab+b^2$; $a^2-2ab+b^2$

Difference of Two Squares

1. $(x-4)(x+4)$

2. $(5x-2)(5x+2)$

3. $(9-5a)(9+5a)$

4. $(3x-10y)(3x+10y)$

Perfect Square Trinomials

5. $(x+6)^2$

6. $(3t+1)^2$

7. $(5x-4)^2$

8. $(x-7y)^2$

Sum and Difference of Two Cubes

9. $(x+5)(x^2-5x+25)$

10. $(a-2)(a^2+2a+4)$

11. $(x-6)(x^2+6x+36)$

12. $(4x-3)(16x^2+12x+9)$

13. $(4y+3)^2$

14. $(4b-5)(4b+5)$

15. $3x(2x-5)^2$

16. $a(3a-8b)(3a+8b)$

6.5 Summary of Factoring

Guidelines for Factoring Polynomials

greatest common factor (GCF)
grouping
a^2-b^2; difference of two squares
a^3-b^3; difference of two cubes
a^3+b^3; sum of two cubes
perfect square
$(a+b)^2$; perfect square trinomial
$(a-b)^2$; perfect square trinomial
grouping; FOIL
completely factored

Factoring Polynomials

1. $5x(x^2-4x+5)$

2. $4t^2(t-6)(t+6)$

3. $-5a(3a+1)^2$

4. $5(x-4)(x^2+4x+16)$

5. $2x^2(4x-1)(3x+2)$

6. $4(x-3)(x+3)(2x+1)$

7. $4ab(2a-3b)(2a+3b)$

8. $(3x^2+5)(4x+3)$

6.6 Solving Equations by Factoring I (Quadratics)

Key Terms

1. zero-product

2. quadratic polynomial

3. zeros

4. quadratic equation

5. standard form

The Zero-Product Property

1. –2, 0

2. 0

3. –1, 4

4. –5, 0, 3

Solving Quadratic Equations

5. –4, 0

6. –3, 3

7. 2, 3

8. $-\dfrac{3}{2}, \dfrac{4}{5}$

9. $-\dfrac{1}{2}, 5$

Applications

10. After 1.5 sec and 4 sec

11. (a) 177.8 ft
(b) About 23 mph
(c) About 23 mph

12. 50×70 pixels

**Understanding Concepts
through Multiple Approaches**

13. **(a)** 36 mph

   (b) 36 mph

6.7 Solving Equations by Factoring II (Higher Degree)

Polynomial with Common Factors

1. $-3(x-4)(x+1)$

2. $2x(x-5)(x-1)$

3. $-\dfrac{1}{3}, 0, \dfrac{1}{2}$

4. $-5, 0, 6$

5. 5 inches

Special Types of Polynomials

6. $(x-3)(x+3)(x^2+9)$

7. $(a^2+1)(a^2+5)$

8. $(r-t)^2(r+t)^2$

9. $(a-2b)(a+2b)(a^2+4b^2)$

10. $x(x-4)(x^2+4x+16)$

7.1 Introduction to Rational Expressions

Key Terms

1. lowest terms

2. probability

3. undefined

4. vertical asymptote

5. rational expression; defined

6. basic principle

Basic Concepts

1. -1

2. $-\dfrac{1}{2}$

3. undefined

4. -1

5. 0

6. $-\dfrac{2}{3}$

7. $-3, 3$

8. None

Simplifying Rational Expressions

9. $-\dfrac{1}{3}$

10. $\dfrac{5}{8}$

11. $\dfrac{4}{t}$

12. $\dfrac{3}{4}$

13. $\dfrac{x-2}{x-1}$

14. $\dfrac{x+5}{2x-1}$

15. $-\dfrac{1}{3}$

16. -1

17. 1

Applications

18. (a)

x (cars/minute)	5	6	7	7.5	7.9	7.99
T (minutes)	$\dfrac{1}{3}$	$\dfrac{1}{2}$	1	2	10	100

(b) As the average traffic rate increases, the time needed to pass through the construction zone increases.

19. (a)

x (months)	0	6	12	36	72
P (thousands)	0.4	1.818	2.235	2.683	2.831

(b) 400 fish

(c) The fish population increased quickly at first but then leveled off.

20. (a) $\dfrac{4}{n}$

(b) $\dfrac{4}{100} = \dfrac{1}{25}$; $\dfrac{4}{1000} = \dfrac{1}{250}$; $\dfrac{4}{10{,}000} = \dfrac{1}{2500}$

(c) As the number of balls increases, the probability of picking the winning ball decreases.

7.2 Multiplication and Division of Rational Expressions

Key Terms

1. numerators; denominators

2. basic principle of fractions

3. reciprocal

4. lowest terms

Review of Multiplication and Division of Fractions

1. $\dfrac{6}{35}$

2. $\dfrac{5}{2}$

3. $\dfrac{1}{24}$

4. $\dfrac{4}{15}$

5. $\dfrac{1}{21}$

6. $\dfrac{3}{8}$

Multiplication of Rational Expressions

7. $\dfrac{4(2x+3)}{x(x+1)}$

8. $\dfrac{3x}{x-5}$

9. $x+3$

10. $\dfrac{x+4}{2(x-2)}$

11. (a) $D=\dfrac{120}{x}$

(b) When $x=0.2$, $D=600$ feet

When $x=0.6$, $D=200$ feet

The car requires 3 times more distance to stop on an icy road than on dry pavement.

Division of Rational Expressions

12. $\dfrac{x+2}{6x}$

13. $\dfrac{x-2}{x^2+1}$

14. $\dfrac{x+3}{x+1}$

7.3 Addition and Subtraction with Like Denominators

Key Terms

1. like terms

2. subtract; numerators; denominator

3. greatest common factor

4. add; numerators; denominator

Review of Addition and Subtraction of Fractions

1. $\dfrac{5}{7}$

2. $\dfrac{3}{4}$

3. 1

4. $\dfrac{2}{3}$

Rational Expressions with Like Denominators

5. $\dfrac{9}{t}$

6. 1

7. $\dfrac{1}{a+5}$

8. $x+2$

9. $\dfrac{7}{ab}$

10. $\dfrac{1}{x-y}$

11. $\dfrac{4}{a-b}$

12. 1

13. $-\dfrac{2y}{2y-5}$

14. $\dfrac{1}{3x-4}$

15. $\dfrac{3x}{3x+4}$

16. $-\dfrac{1}{2}$

17. (a) $\dfrac{13}{n}$
    (b) There are 13 chances in n that a defective calculator is chosen.

7.4 Addition and Subtraction with Unlike Denominators

Key Terms

1. prime factorization method

2. least common denominator

3. common multiple

4. listing method

5. least common multiple

Finding Least Common Multiples

1. $12a^2$

2. z^2+z

3. $(x-4)(x+5)$

4. $(x-1)(x-2)(x+3)$

5. $3x^2(x-5)(x+1)$

Review of Fractions with Unlike Denominators

6. $\dfrac{1}{36}$

7. $\dfrac{31}{35}$

Rational Expressions with Unlike Denominators

8. $\dfrac{15x}{9x^2}$

9. $\dfrac{2(x+y)}{x^2-y^2}=\dfrac{2x+2y}{x^2-y^2}$

10. $\dfrac{8y+5}{18y^2}$

11. $\dfrac{2x}{(x+2)(x-2)}$

12. $\dfrac{2x+3}{(x+3)^2}$

13. $\dfrac{-a^2+4a+8}{a(a+2)}$

14. $\dfrac{3x+1}{(x-1)(x+1)}$

15. $\dfrac{x^2-2x-1}{x(x-1)(x+1)}$

16. $\dfrac{3}{x}$

17. $\dfrac{4x-8}{(x-5)(x+1)} = \dfrac{4(x-2)}{(x-5)(x+1)}$

3. $\dfrac{xy}{2}$

4. $3(z+2)$

5. $\dfrac{y-x}{y+x}$

6. $\dfrac{a^2+3}{a^2-3}$

7. $\dfrac{2x-1}{x-3}$

8. $\dfrac{ab}{b-a}$

9. $\dfrac{1}{x^2}$

10. $\dfrac{x}{5x+2}$

11. $\dfrac{3xy}{y+x}$

7.5 Complex Fractions

Key Terms

1. complex fraction

2. least common denominator

3. reciprocal

4. basic complex fraction

Simplifying Complex Fractions

1. $\dfrac{3}{4}$

2. $\dfrac{7}{10}$

7.6 Rational Equations and Formulas

Key Terms

1. rational equation

2. algebraically

3. least common denominator;
   least common denominator;
   term; extraneous solution

4. visually

5. basic rational equation

6. numerically

Solving Rational Equations

1. $\dfrac{15}{4}$

2. $-\dfrac{2}{3}$

3. $-\dfrac{1}{2},\ 4$

4. $\dfrac{7}{5}$

5. -10

6. $\dfrac{5}{2}$

7. No solutions

Rational Expressions and Equations

8. Expression; $\dfrac{x^2-2}{x+2};\ \dfrac{7}{3}$

9. Equation; $x = \dfrac{1}{3}$

Graphical and Numerical Solutions

10. $-1,\ 3$

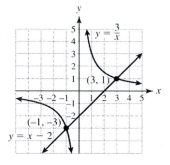

$-1,\ 3$

x	-3	-2	-1	0	1	2	3
$y_1 = \frac{3}{x}$	-1	$-\frac{3}{2}$	-3	$--$	3	$\frac{3}{2}$	1
$y_2 = x - 2$	-5	-4	-3	-2	-1	0	1

Solving a Formula for a Variable

11. (a) 150 miles

(b) $t = \dfrac{d}{r}$

(c) 3 hours, 20 minutes

12. $h = \dfrac{2A}{b}$

13. $T = \dfrac{PV}{nR}$

14. $b_2 = \dfrac{2A - b_1 h}{h}$ or $b_2 = \dfrac{2A}{h} - b_1$

Applications

15. $\dfrac{6}{5}$ hour, or 1 hr 12 min

16. $\dfrac{8}{5}$ hour, or 1 hr 36 min

Understanding Concepts through Multiple Approaches

17. (a) $x = 4$

(b)

x	-2	0	2	4	6
$y = \dfrac{2}{x-3}$	$-\dfrac{2}{5}$	$-\dfrac{2}{3}$	-2	2	$\dfrac{2}{3}$
$y = 2$	2	2	2	2	2

(c)

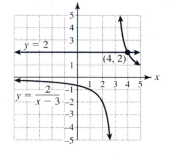

7.7   Proportions and Variation

Key Terms

1. inversely proportional;
    varies inversely

2. ratio

3. directly proportional;
    varies directly

4. proportion

5. varies jointly

6. constant of proportionality;
    constant of variation

Proportions

1. 1.75 in.

2. 16 feet

Direct Variation

3. $y = \dfrac{20}{3}$

4. $1652

5. **(a)** No
    (b) No

Inverse Variation

6. $y = 5$

7. **(a)** No
    (b) Yes
    (c) $y = 3$

Analyzing Data

8. Inverse

9. Neither

10. Direct

11. 120 lb

8.1 Functions and Their Representations

Key Terms

1. relation

2. function; domain; range

3. function notation; input; output; name of the function

4. nonlinear

5. dependent; independent

6. vertical line test

Representations of a Function

1. $f(-3) = -15$

2. $f(4) = -4$

3. $f(3) = 4$

4. (a) Multiply the input 3 by 0.05 to obtain 0.15. The sales tax on a $3.00 purchase is $0.15.

   (b) From the table, $f(3) = \$0.15$.

   (c) $f(3) = 0.05(3) = \$0.15$

   (d) From the graph, $f(3) = \$0.15$.

Sales Tax of 5%

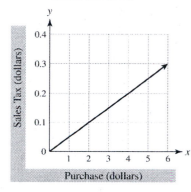

4. (e) $f(3) = \$0.15$

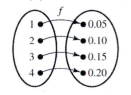

5. (a) $S(2010) = 1600$ billion

   In 2010, there were 1600 billion World Wide Web searches.

   (b) 225 billion

6. (a) $f(x) = x^2 - 4$

   (b)

x	-2	-1	0	1	2
$f(x)$	0	-3	-4	-3	0

   (c)

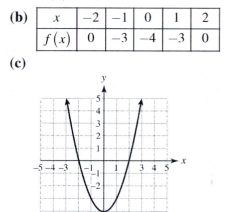

Definition of a Function

7. (a) $\{(1,68), (3,77), (5,84), (7,92)\}$

   (b) $D = \{1, 3, 5, 7\}$

   $R = \{68, 77, 84, 92\}$

   (c) The longer the study time, the higher the exam score.

8. D: All real numbers; $R: y \geq 3$

9. D: All real numbers; $R: y = 2$

10. D: All real numbers

11. $D: x \neq 0$

12. $D: x \geq 4$

Identifying a Function

13. Yes

14. No

15. Yes

16. No

17. No

Graphing Calculators (Optional)

18.

[-5, 10, 0.75] by [-50, 50, 5]

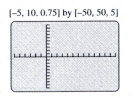

19.

8.2   Linear Functions

Key Terms

1.   linear function

2.   rate of change; y-intercept

3.   nonlinear function

4.   midpoint

5.   constant function

Basic Concepts

1.   yes; $m = -\dfrac{1}{2}, b = 3$

2.   yes; $m = 4, b = 0$

3.   not a linear function

4.   yes; $f(x) = 2x - 1$

5.   not a linear function

6.   yes; $f(x) = -2$

7.   yes; $f(x) = -x$

Representations of Linear Functions

8. $f(-4) = 2$

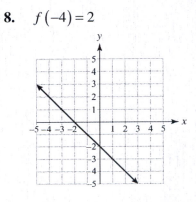

9.

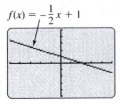

$f(x) = -\dfrac{1}{2}x + 1$

10.   **(a)** Multiply the input x by 2 and subtract 1 to obtain the output.

(b)

x	-1	0	1
$f(x)$	-3	-1	1

(c)

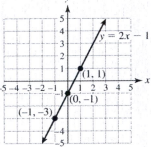

Modeling Data with Linear Functions

11. (a) $C(x) = 1200 + 75x$

    (b) $12,450$; The total cost to manufacture 150 items is $12,450$.

12. (a) Recommended Cumulative Weight Loss

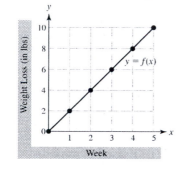

    (b) 8 lbs, 2 lbs

    (c) $f(x) = 2x$

    (d) 24 lbs

13. (a) The pace of the athlete appears to be constant at 8 miles per hour.

    (b) $f(x) = 8$

    (c)

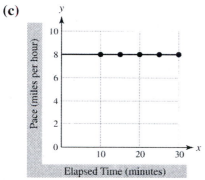

The Midpoint Formula (Optional)

14. $\left(\dfrac{1}{2}, -2\right)$

15. (a) 4.2 per 1000 people

    (b) No

16. (a) $f(x) = -x + 1$

    (b) $f(0) = 1$; yes

    (c) $\left(-\dfrac{1}{2}, \dfrac{3}{2}\right)$

8.3   Compound Inequalities

Key Terms

1. compound

2. union; $A \cup B$

3. set-builder

4. three-part

5. interval

6. intersection; $A \cap B$

Basic Concepts

1. no; yes

2. yes; no

Symbolic Solutions and Number Lines

3. $\{x \mid -1 \le x < 2\}$

4. $\{x \mid 1 < x \le 2\}$

5. $\{x \mid -2 < x < 6\}$

6. $\{m \mid -5 < m \le 4\}$

7. from 1.4 miles to 2.1 miles

8. all real numbers

Numerical and Graphical Solutions

9. from 1995 to 2000

Interval Notation

10. $(-3,4]$

11. $\left[-\dfrac{1}{2},\infty\right)$

12. $(-\infty,-3)\cup[1,\infty)$

13. $[-5,0)$

14. $(-\infty,-2]\cup\left(\dfrac{5}{2},\infty\right)$

15. $(2,\infty)$

8.4   Other Functions and Their Properties

Key Terms

1. absolute value function

2. linear function

3. quadratic function

4. cubic function

5. rational function

Expressing Domain and Range in Interval Notation

1. $(-\infty,\infty)$

2. $[4,\infty)$

3. $(-\infty,0)\cup(0,\infty)$

4. $D=[-2,\infty);\ R=[0,\infty)$

Polynomial Functions

5. Yes; quadratic; 2

6. No

7. No

8. Yes; linear; 1

9. -3

10. 6

11. 0

12. (a)   180 bpm
     (b)   79 bpm

Rational Functions (Optional)

13. $(-\infty,-7)\cup(-7,\infty)$

14. $(-\infty,1)\cup(1,4)\cup(4,\infty)$

15. $(-\infty,-2)\cup(-2,0)\cup(0,\infty)$

16. $D=(-\infty,-3)\cup(-3,\infty)$

17. (a) $f(-2)$ is undefined
         $f(-1)=1$
         $f(1)=-\dfrac{1}{3}$
     (b) $f(-2)$ is undefined
         $f(-1)=1$
         $f(1)=-\dfrac{1}{3}$

17. (c) $f(-2)$ is undefined

$$f(-1) = 1$$

$$f(1) = -\frac{1}{3}$$

18. (a) $f(85) = 6.25\%$

The least active 85% of the population contributes only 6.25% of the postings.

(b) $x \approx 91\%$

The least active 91% of the population contributes only 10% of the postings.

Operations on Functions

19. 3

20. -36

21. Undefined

22. $-\dfrac{1}{4}$

23. $5x + 6$

24. $3x - 8$

25. $4x^2 + 27x - 7$

26. $\dfrac{4x-1}{x+7}$

8.5 Absolute Value Equations and Inequalities

Key Terms

1. $|ax+b| = k$

2. $|ax+b| = |cx+d|$

3. $|ax+b| < k$

4. $|ax+b| > k$

Absolute Value Equations

1. 0

2. $-3, 3$

3. $-\dfrac{13}{3}, 5$

4. $-\dfrac{1}{2}, \dfrac{3}{2}$

5. $-\dfrac{9}{2}, -\dfrac{7}{2}$

6. No solutions

7. $\dfrac{1}{4}$

8. $-1, \dfrac{1}{2}$

Absolute Value Inequalities

9. $-2, 3$

10. $[-2, 3]$

11. $(-\infty, -2) \cup (3, \infty)$

12. $(-\infty, -9] \cup [6, \infty)$

13. $|w - 32.4| \le 0.03$

14. $43 \le T \le 97$; Average monthly temperatures range between $43°F$ and $97°F$.

15. No solutions

16. All real numbers

**Understanding Concepts
through Multiple Approaches**

17. **(a)** $\{x \,|\, x \le -3 \text{ or } x \ge 0\}$

   (b)

x	-4	-3	-2	-1	0	1
$|2x+3|$	5	3	1	1	3	5

   (c)

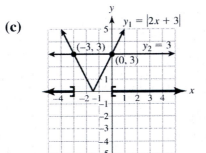

9.1 Systems of Linear Equations in Three Variables

Key Terms

1. linear system in three variables; ordered triple

2. eliminate; eliminate; substitute; solve; substitute; solution

Basic Concepts

1. $(-2,1,2)$

2. $x + y + z = 180$
$x - y - z = 20$
$x \quad - z = 70$

3. $2a + b + c = 39$
$2a + 3b \quad = 49$
$a + b + 4c = 67$

Solving Linear Systems with Substitution and Elimination

4. $(-2,-1,3)$

5. $(3,-3,0)$

6. 150 child tickets
200 student tickets
400 adult tickets

Modeling Data

7. $a = 20,\ b = -5,\ c = 70$

Systems of Equations with No Solutions

8. No solutions

Systems of Equations with Infinitely Many Solutions

9. $(2z+1,-1,z)$

9.2 Matrix Solutions of Linear Equations

Key Terms

1. dimension

2. matrix

3. Gauss-Jordan elimination; matrix row transformations

4. square

5. reduced row-echelon form

6. main diagonal; augmented

7. element

Representing Systems of Linear Equations with Matrices

1. $\begin{bmatrix} 1 & 3 & | & 13 \\ 5 & -3 & | & -25 \end{bmatrix}$
2×3

2. $\begin{bmatrix} 1 & -1 & 2 & | & -2 \\ 2 & -4 & 3 & | & 1 \\ -2 & 2 & -3 & | & 2 \end{bmatrix}$
3×4

Matrices and Social Networks

3. $\begin{bmatrix} 0 & 0 & 1 & 1 \\ 1 & 0 & 0 & 0 \\ 1 & 1 & 0 & 0 \\ 1 & 1 & 0 & 0 \end{bmatrix}$

Gauss-Jordan Elimination

4. $(0,-5)$

5. $(-2,4,5)$

6. $\begin{bmatrix} 1 & -3 & | & -4 \\ 2 & -1 & | & -3 \end{bmatrix}$; $(-1,1)$

7. $\begin{bmatrix} 1 & 2 & 1 & | & 5 \\ 1 & 0 & -2 & | & 4 \\ 2 & 3 & 0 & | & -5 \end{bmatrix}$; $(2,-3,-1)$

8. $3000 at 4%
   $4000 at 7%
   $8000 at 8%

Using Technology to Solve Systems of Linear Equations (Optional)

9. $\left(\dfrac{1}{2}, -\dfrac{1}{2} \right)$

10. $(-4,0,3)$

11. (a) $a + 15b + 2c = 215$
       $a + 10b + 4c = 450$
       $a + 25b + c = 75$
    (b) $a = 40$, $b = -3$, $c = 110$
    (c) $355,000

Calculation of Determinants

1. -5

2. -23

3. -36

4. 10

5. -36

6. 10

Area of Regions

7. 6.5 square miles

Cramer's Rule

8. $(3,-2)$

9. $\left(-\dfrac{13}{2}, -\dfrac{17}{2} \right)$

9.3   Determinants

Key Terms

1. determinant; b

2. determinants; determinants; expansion by minors

3. b_3, a_2, c_1, c_2; minors

4. c_1, c_2, a_2; Cramer's rule

10.1 Radical Expressions and Functions

Key Terms

1. radical sign
2. cube root
3. principal nth root
4. radical expression
5. square root
6. nth root
7. index; odd root; even root
8. radicand

Radical Notation

1. ± 12
2. 8
3. 0.6
4. $\dfrac{4}{5}$
5. $|d|$
6. 4.359
7. (a) about 3 miles
   (b) about 119 miles
8. 3
9. -4
10. $\dfrac{1}{2}$

11. a^4
12. 2.29
13. 5
14. -3
15. Not a real number
16. -2
17. $|z|$
18. $|x-2|$
19. $|x+3|$

The Square Root Function

20. $f(1)=1$
    $f(-2)=\sqrt{-11}$; Not a real number
21. $f(1)=2\sqrt{3}$
    $f(-2)=3$
22. (a) about 2.7 seconds
    (b) No, the hang time does not triple.
23. (a) $[3,\infty)$
    (b) Graph is shifted three units to the right.

24. $[-2,\infty)$

25. $(-\infty,\infty)$

26. (a) $264 thousand
$373 thousand
$457 thousand
(b) $84 thousand
$109 thousand
There is more revenue gained from
4 to 8 employees than there is from
8 to 12 employees.

The Cube Root Function

27. $f(1)=1$
$f(-3)=\sqrt[3]{17}$

28. $f(1)=2$
$f(-3)=0$

10.2 Rational Exponents

Key Terms

1. $\sqrt[n]{a}$

2. $\sqrt[n]{a^m}=\left(\sqrt[n]{a}\right)^m$

3. $\dfrac{1}{a^{m/n}}=\dfrac{1}{\sqrt[n]{a^m}}$

4. a^{p+q}

5. $\dfrac{1}{a^p}$

6. $\left(\dfrac{b}{a}\right)^p$

7. a^{p-q}

8. a^{pq}

9. $a^p b^p$

10. $\dfrac{a^p}{b^p}$

Basic Concepts

1. $\sqrt{49}=7$

2. $\sqrt[3]{15}\approx 2.47$

3. $\sqrt[4]{x}$

4. $\sqrt{3z}$

5. $(-64)^{2/3}=\left(\sqrt[3]{-64}\right)^2=16$

6. $(32)^{3/5}=\left(\sqrt[5]{32}\right)^3=8$

7. $A(20)\approx 27\%$
$A(60)\approx 44\%$

After 20 seconds about 27% of the
viewers have abandoned the online
video, and after 60 seconds just under
half, or 44%, of the viewers have
abandoned the online video.

8. $16^{-3/4}=\dfrac{1}{\left(\sqrt[4]{16}\right)^3}=\dfrac{1}{8}$

9. $-27^{-2/3}=-\dfrac{1}{\left(\sqrt[3]{27}\right)^2}=-\dfrac{1}{9}$

10. $x^{5/6}$

11. $a^{-7/2}$

12. $(x+4)^{3/4}$

13. $\left(s^4-t^4\right)^{1/4}$

14. Approximately 0.86 step per second

Properties of Rational Exponents

15. $x^{3/4}$

16. $2^{5/4} x^{3/2}$

17. $\dfrac{3}{x^{1/6}}$

18. $\dfrac{y}{7}$

19. $(z+3)^{1/8}$

20. y^3

21. $\dfrac{b^{1/4}}{a^{1/3}}$

22. $x^{1/2} + x$

10.3 Simplifying Radical Expressions

Key Terms

1. perfect nth power

2. $\dfrac{\sqrt[n]{a}}{\sqrt[n]{b}}$

3. perfect cube

4. $\sqrt[n]{a \cdot b}$

5. perfect square

Product Rule for Radical Expressions

1. 10

2. -5

3. $\dfrac{1}{2}$

4. x^4

5. $\sqrt[3]{28 y^2}$

6. $\sqrt{6ab}$

7. 3

8. $5\sqrt{3}$

9. $3\sqrt[3]{3}$

10. $2\sqrt{6}$

11. $3\sqrt[4]{2}$

12. (a) $L = 1.5\sqrt{R}$
    (b) 45 mph

13. $4 y^3$

14. $2x^2 \sqrt{2x}$

15. $-3ab^2 \sqrt[3]{2a}$

16. $4m\sqrt[3]{n}$

17. $6^{5/6} = \sqrt[6]{7776}$

18. 3

19. $b^{9/20} = \sqrt[20]{b^9}$

Quotient Rule for Radical Expressions

20. $\dfrac{\sqrt[3]{3}}{2}$

21. $\dfrac{\sqrt[4]{x}}{3}$

22. $\dfrac{5}{t}$

23. 3

24. $\dfrac{1}{5}$

25. m

26. $\dfrac{y\sqrt[4]{y}}{2x}$

27. $\dfrac{6x^2}{5}$

28. $\sqrt{x^2-16}$

29. $\sqrt[3]{x+1}$

10.4 Operations on Radical Expressions

Key Terms

1. like radicals

2. rationalizing the denominator

3. conjugate

Addition and Subtraction

1. $13\sqrt{7}$

2. $5\sqrt[3]{5}$

3. $3+2\sqrt{3}$; not possible

4. $\sqrt{6}+\sqrt{10}$; not possible

5. $5\sqrt{3},\ 4\sqrt{3}$

6. $\sqrt{8}=2\sqrt{2},\ \sqrt{4}=2$; not possible

7. $14\sqrt[3]{3},\ 6\sqrt[3]{3}$

8. $7\sqrt{2}$

9. $3\sqrt[3]{3}$

10. $10\sqrt{2}$

11. $5\sqrt[4]{2}$

12. $-11\sqrt{x}$

13. $27\sqrt{5b}$

14. $4\sqrt{5}$

15. $\sqrt[3]{7}+\sqrt[3]{10}$

16. $3\sqrt{t}+\sqrt[3]{t}$

17. $4\sqrt[3]{m^2n}$

18. $2y\sqrt{y}$

19. $\dfrac{\sqrt[3]{7x}}{4}$

20. **(a)** $425\sqrt{x}+3300$
    (b) $\$88,300$

21. $\dfrac{13\sqrt{2}}{12}$

22. $\left(3xy^2-2\right)\sqrt[4]{x^2y}$

23. $(t-4)\sqrt[3]{t}$

24. $14\sqrt{3}$ feet ≈ 24.25 feet

Multiplication

25. $a + 4\sqrt{a} - 12$

26. -1

Rationalizing the Denominator

27. $\dfrac{\sqrt{3}}{3}$

28. $\dfrac{\sqrt{2}}{7}$

29. $\dfrac{\sqrt{2y}}{10}$

30. $y\sqrt{x}$

31. $x\sqrt{3}$

32. $3\sqrt{5} - 6$

33. $\dfrac{9 - 5\sqrt{3}}{6}$

34. $\dfrac{x + 4\sqrt{x}}{x - 16}$

35. $\dfrac{4\sqrt[3]{y^2}}{y}$

10.5 More Radical Functions

Key Terms

1. power function; root function

Root Functions

1. $f(-5) = \sqrt{-1}$; Not a real number
    $f(5) = \sqrt{9} = 3$

2. $f(-5) = \sqrt[3]{8} = 2$
    $f(30) = \sqrt[3]{-27} = -3$

3. $f(1) = \sqrt[4]{-4}$; Not a real number
    $f(21) = \sqrt[4]{16} = 2$

4. $f(-243) = \sqrt[5]{-243} = -3$
    $f(1) = \sqrt[5]{1} = 1$

Power Functions

5. Not possible

6. 9

7. (a) 3181 square inches
    (b) 138 square inches

8. Larger exponents result in graphs that increase (rise) faster for $x > 1$. Smaller exponents result in graphs that have larger y-values for $0 < x < 1$.

Modeling with Power Functions (Optional)

9. (a) $k \approx 0.85$
    (b) About 0.717 m

10.6 Equations Involving Radical Expressions

Key Terms

1. extraneous solutions

2. Pythagorean theorem

3. distance

Solving Radical Equations

1. 10

2. 14

3. 2

4. 5

5. -4

6. Approximately 5.2 pounds

7. $x \approx -2.52, 1.94$

The Distance Formula

8. 16 feet

9. $11\sqrt{2}$

Solving the Equations $x^n = k$

10. -3

11. No real solutions

12. 2, 8

13. (a) $v(W) = \sqrt[3]{\dfrac{W}{2.8}}$

(b) 8.1 mph

Understanding Concepts through Multiple Approaches

14. (a) 33

(b) 33

(c) 33

10.7 Complex Numbers

Key Terms

1. complex; standard form; real; imaginary

2. imaginary

3. complex conjugate

4. pure imaginary

5. imaginary

6. real

Basic Concepts

1. $4i$

2. $i\sqrt{5}$

3. $5i\sqrt{2}$

Addition, Subtraction, and Multiplication

4. $9+9i$

5. $-10+2i$

6. $2+11i$

7. 40

Powers of i

8. 1

9. i

10. -1

Complex Conjugates and Division

11. $4-i$

12. $-5+2i$

13. $-6i$

14. 3

15. $-\dfrac{9}{10}-\dfrac{13}{10}i$

16. $3i$

11.1 Quadratic Functions and Their Graphs

Key Terms

1. vertex

2. reflection

3. axis of symmetry

4. vertical shift

5. quadratic function

6. (a) vertex; axis of symmetry
   (b) upward; downward
   (c) wider
   (d) narrower

Graphs of Quadratic Functions

1. $(2,0)$; $x = 2$; downward

2. $(0,-4)$; $x = 0$; upward

3. $(-1,2)$

4. $(0,-2)$; $x = 0$

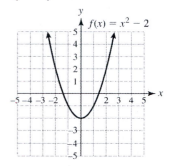

5. $(-4,0)$; $x = -4$

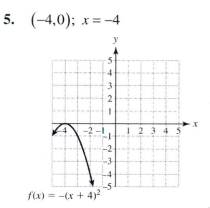

$f(x) = -(x + 4)^2$

6. $(-1,1)$; $x = -1$

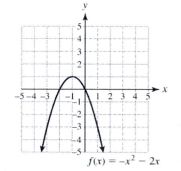

$f(x) = -x^2 - 2x$

Min-Max Applications

7. (a) $f(x) = x(30 - 0.50x)$
   (b) 30 shirts

8. $\dfrac{1}{4}$

9. (a) 4 feet
   (b) 148 feet

10. (a) The revenue increases at first, reaches a maximum (the vertex), and then decreases.
    (b) $784; 14 rooms
    (c) $f(x) = x(112 - 4x)$
    (c) $784; 14 rooms

Basic Transformations of $y = ax^2$

11.   Both graphs are parabolas. The graph of g opens downward and is wider than the graph of f.

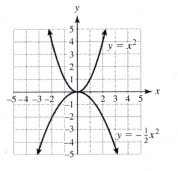

Transformations of $y = ax^2 + bx + c$
(Optional)

12.   **(a)**   opens upward
          wider

   (b) $x = 2; \left(2, -\dfrac{9}{2}\right)$

   (c) y-intercept: $-\dfrac{5}{2}$
          x-intercepts: $-1, 5$

   (d)

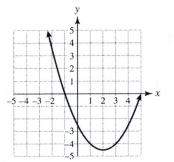

11.2 Parabolas and Modeling

Key Terms

1. $y = (x - h)^2$

2.   translations

3. $y = x^2 + k$

4.   vertex; $y = a(x - h)^2 + k$;
        upward; downward

5. $y = (x + h)^2$

6.   completing the square

7. $y = x^2 - k$

Vertical and Horizontal Translations

1.

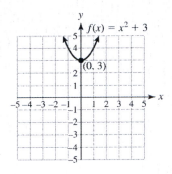

2.

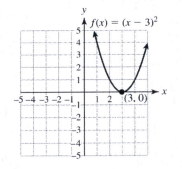

3.

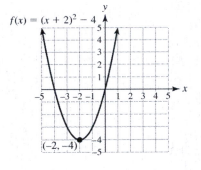

Vertex Form

4. The graph of $y = f(x)$ is translated 1 unit right and 3 units downward. The graph opens upward and is narrower.

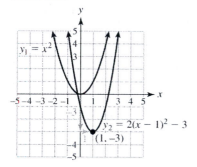

5. The graph of $y = f(x)$ is translated 2 units left and 4 units upward. The graph opens downward and is wider.

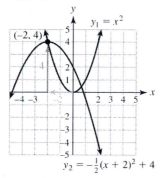

6. $y = -(x+3)^2 - 1$
$y = -x^2 - 6x - 10$

7. $(1, -3)$
$y = 2(x-1)^2 - 3$

8. $y = (x-3)^2 - 9$
$(3, -9)$

9. $y = (x+2)^2 + 1$
$(-2, 1)$

10. $y = \left(x + \dfrac{5}{2}\right)^2 - \dfrac{33}{4}$
$\left(-\dfrac{5}{2}, -\dfrac{33}{4}\right)$

11. $y = -2\left(x - \dfrac{3}{2}\right)^2 + \dfrac{1}{2}$
$\left(\dfrac{3}{2}, \dfrac{1}{2}\right)$

Modeling with Quadratic Functions (Optional)

12. $a = -\dfrac{1}{3}$

11.3 Quadratic Equations

Key Terms

1. quadratic equation

2. square root property

Basics of Quadratic Equations

1. No real solutions

2. $2, -3$

3. 1

The Square Root Property

4. $\pm 2\sqrt{2}$

5. $\pm \dfrac{5}{2}$

6. $-7, 1$

7. $\sqrt{5}$ sec ≈ 2.2 sec

Completing the Square

8. 16

9. $-2 \pm \sqrt{7}$

10. $\dfrac{5 \pm \sqrt{57}}{4}$

Solving an Equation for a Variable

11. $y = \pm \dfrac{\sqrt{x-1}}{3}$

12. $r = \sqrt{\dfrac{V}{\pi h}}$

Applications of Quadratic Equations

13. 20 mph

14. 153.5 million

**Understanding Concepts
through Multiple Approaches**

15. (a) 2, 3

    (b) 2, 3

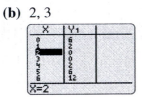

    (c) 2, 3

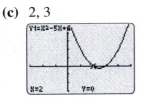

11.4 The Quadratic Formula

Key Terms

1. one

2. quadratic formula

3. no

4. discriminant

5. two

Solving Quadratic Equations

1. $\dfrac{7 \pm \sqrt{33}}{8}$

2. $-\dfrac{4}{3}$

3. No real solutions

4. 58.8 mph ≈ 60 mph

The Discriminant

5. (a) $a > 0$
   (b) $-2, 4$
   (c) positive

6. one real solution; $-\dfrac{3}{5}$

**Quadratic Equations
Having Complex Solutions**

7. $\pm i\sqrt{7}$

8. $-\dfrac{1}{4} \pm i\dfrac{\sqrt{31}}{4}$

9. $\dfrac{3}{2} \pm i\dfrac{\sqrt{15}}{2}$

10. $3 \pm 4i$

11.5 Quadratic Inequalities

Key Terms

1. quadratic inequality

2. test value

3. $p < x < q$
   $x < p$ or $x > q$

Basic Concepts

1. yes

2. no

Graphical and Numerical Solutions

3. $[-3, 2]$

x	$y = x^2 + x - 6$
-4	6
-3	0
-2	-4
-1	-6
0	-6
1	-4
2	0
3	6

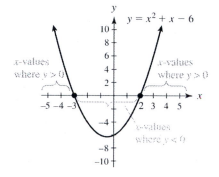

4. $(-\infty, -4] \cup [4, \infty)$

5. No real solutions

6. $(-\infty, \infty)$

7. $(-\infty, 2) \cup (2, \infty)$

8.

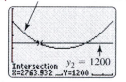

8. (continued)

$y_1 = 0.00004x^2 - 0.4x + 2000$

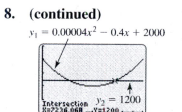

Symbolic Solutions

9. $[2, 5]$

10. $\left(-\infty, -\dfrac{1}{3}\right) \cup \left(\dfrac{1}{2}, \infty\right)$

11. 20 feet or more

11.6 Equations in Quadratic Form

Higher Degree Polynomial Equations

1. $-\sqrt[3]{\dfrac{3}{4}}, 1$

Equations Having Rational Exponents

2. $-\dfrac{3}{4}, \dfrac{2}{3}$

3. $-\dfrac{1}{8}, -8$

Equations Having Complex Solutions

4. $\pm 3, \pm 3i$

5. $-\dfrac{1}{4} \pm i\dfrac{\sqrt{7}}{4}$

12.1 Composite and Inverse Functions

Key Terms

1. vertical line test;
   horizontal line test

2. one-to-one

3. horizontal line test

4. composite function;
   composition

5. inverse function

Composition of Functions

1. 10; $(g \circ f)(x) = 2x^3 - 6$

2. 19; $(g \circ f)(x) = 9x^2 - 6x - 5$

3. undefined; $(g \circ f)(x) = \dfrac{1}{\sqrt{2x - 2}}$

4. 0

5. 2

6. 3

7. −2

One-to-One Functions

8. One-to-one

9. Not one-to-one

Inverse Functions

10. Divide x by −2.
    $$f(x) = -2x$$
    $$g(x) = \frac{x}{-2} = -\frac{x}{2}$$

11. Add 15 to x and then divide the result by 2.
    $$f(x) = 2x - 15$$
    $$g(x) = \frac{x + 15}{2}$$

12. Answers may vary.
    $$f\left(f^{-1}(x)\right) = f^{-1}\left(f(x)\right) = x$$

13. $f^{-1}(x) = \dfrac{x + 1}{4}$

14. $f^{-1}(x) = \sqrt[3]{x} + 1$

Tables and Graphs of Inverse Functions

15.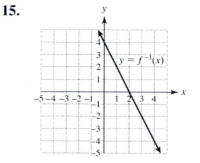

16. (a) Yes

    (b)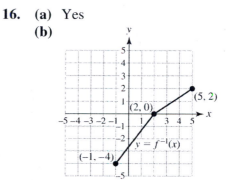

12.2 Exponential Functions

Key Terms

1. compound interest; growth factor

2. natural exponential

3. percent change

4. exponential; base; coefficient

5. continuous growth

6. growth factor; decay factor

Basic Concepts

1. 100

2. 36

3. $\dfrac{1}{8}$

Graphs of Exponential Functions

4. (a)

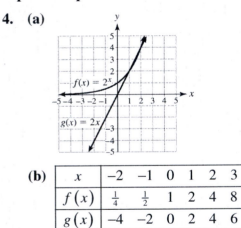

(b)

x	-2	-1	0	1	2	3
$f(x)$	$\frac{1}{4}$	$\frac{1}{2}$	1	2	4	8
$g(x)$	-4	-2	0	2	4	6

5. Exponential; $f(x) = \left(\dfrac{1}{5}\right)^{x} = (5)^{-x}$

6. Linear; $f(x) = 2x - 1$

7. Exponential; $f(x) = \dfrac{1}{4}(2)^{x}$

Percent Change and Exponential Functions

8. (a) 25%
    (b) -40%

9. (a) $1250
    (b) $6250
    (c) 1.25

10. (a) $B(x) = 80{,}000(0.90)^{x}$
    (b) $52{,}488$; After 4 hours, $52{,}488$ bacteria remain.

Compound Interest

11. $66,338.39; 1.09

12. $3177.17

Models Involving Exponential Functions

13. (a) ≈ 0.63; There is a 63% chance that no vehicle will enter the intersection during any particular 10-second interval.
    (b) Exponential decay

14. (a) ≈ 0.09; There is a 9% chance that no tree is growing within any particular circle of radius 10 feet.
    (b) $P(x)$ gets closer to zero.

The Natural Exponential Function

15. 582 thousand

12.3 Logarithmic Functions

Key Terms

1. common logarithm

2. logarithmic function with base a

3. natural logarithm

4. logarithm with base a

5. common logarithmic function

The Common Logarithmic Function

1. 4

2. Not possible

3. 3

4. -2

5. 1

6. 1.929

The Inverse of the Common Logarithmic Function

7. e

8. $x^2 - 1$

9. 3

10. $8x$

11. The graph is moved to the left 3 units.

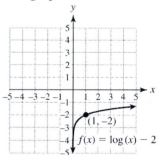

12. The graph is moved down 2 units.

13. 130 dB

Logarithms with Other Bases

14. -3

15. 2

16. 4.61

17. -1.61

18. 3

19. -3

20. 0

21. -4

22. $12x$

23. 1.5

24. $2x$, for $x > 0$

25. $3x + 1$, for $x > -\dfrac{1}{3}$

26. (a) Approximately 3500 ft
    (b) No. Answers will vary.

12.4 Properties of Logarithms

Key Terms

1. $\log_a m + \log_a n$

2. $\log_a m - \log_a n$

3. $r \log_a m$

4. $\dfrac{\log x}{\log a}; \dfrac{\ln x}{\ln a}$

Basic Properties

1. $\log 3 + \log 11$

2. $\ln 7 + \ln x$

3. $\ln x + \ln x + \ln x$

4. $\log_2 12$

5. $\log xy^2$

6. $\ln 8x^2$

7. $\ln 5 - \ln 2$

8. $\log x - \log 4y$

9. $\log_2 y^3 - \log_2 z$

10. $\ln 4$

11. $\log_5 y^3$

12. $\log 6$

13. $4\ln 3$

14. $(x+2)\log(1.4)$

15. $2x\log_3 7$

16. $\ln x^6$

17. $\log x^{5/2}$

18. $\ln x + \dfrac{1}{3}\ln y - 2\ln z$

19. $\dfrac{1}{2}\log y - \dfrac{1}{2}\log x - \dfrac{1}{2}\log z$

20. 1.8

21. 3.2

22. 0.2

Change of Base Formula

23. 2.262

12.5 Exponential and Logarithmic Equations

Exponential Equations and Models

1. 2.26

2. 3.91

3. 5.32

4. 3.11

5. 0.69

6. 4

7. −2

8. 0.19

9. (a) 2
    (b) 2.01

10. 3.66

11. (a) ≈ 0.617; After 6 months, about 61.7% of the robins were still alive.
    (b) After about 1 year

Logarithmic Equations and Models

12. 1000

13. 223.24

14. 35

15. 4

16. About 25,000 pounds

Understanding Concepts through Multiple Approaches

17. (a) ≈ 3.98
    (b) ≈ 4

13.1 Parabolas and Circles

Key Terms

1. conic sections

2. standard equation of a circle

3. circle; radius; center

Parabolas with Horizontal Axes of Symmetry

1. Vertex: $(0,0)$

   Axis of symmetry: $y = 0$

   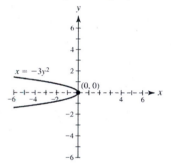

2. Vertex: $(-3,-2)$

   Axis of symmetry: $y = -2$

   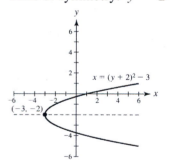

3. Vertex: $(0,-3)$

   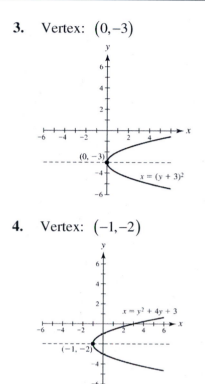

4. Vertex: $(-1,-2)$

   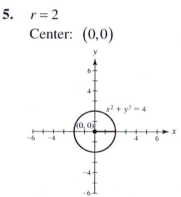

Equations of Circles

5. $r = 2$

   Center: $(0,0)$

6. $r = 3$

Center: $(2, -3)$

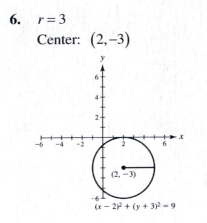

$(x - 2)^2 + (y + 3)^2 = 9$

7. Center: $(1, -3)$

$r = 4$

13.2 Ellipses and Hyperbolas

Key Terms

1. hyperbola; focus

2. asymptotes

3. center

4. transverse axis

5. ellipse; focus

6. branches

7. vertices

Equations of Ellipses

1.

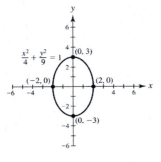

$\dfrac{x^2}{4} + \dfrac{y^2}{9} = 1$

2.

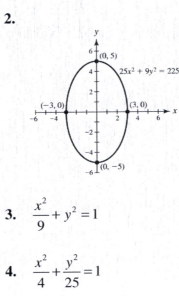

$25x^2 + 9y^2 = 225$

3. $\dfrac{x^2}{9} + y^2 = 1$

4. $\dfrac{x^2}{4} + \dfrac{y^2}{25} = 1$

Equations of Hyperbolas

5.

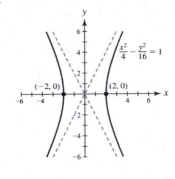

$\dfrac{x^2}{4} - \dfrac{y^2}{16} = 1$

6. $\dfrac{y^2}{9} - \dfrac{x^2}{9} = 1$

13.3 Nonlinear Systems of Equations and Inequalities

Key Terms

1. nonlinear systems of equations

2. nonlinear systems of inequalities

3. method of substitution

Solving Nonlinear Systems of Equations

1. $(2,6), (-2,-6)$

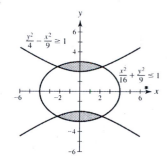

2. $(3,1), (6,-2)$

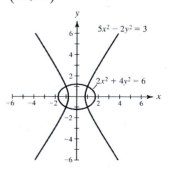

3. $(\pm1, \pm1)$

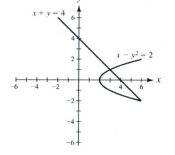

4. $r = 1.5$ inches
$h \approx 8.49$ inches

Solving Nonlinear Systems of Inequalities

5.

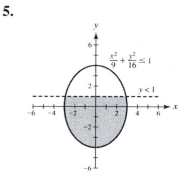

6.

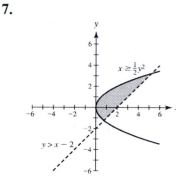

7.

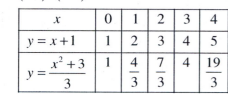

Understanding Concepts through Multiple Approaches

8. **(a)** $(0,1), (3,4)$

(b) $(0,1), (3,4)$

x	0	1	2	3	4
$y = x+1$	1	2	3	4	5
$y = \dfrac{x^2+3}{3}$	1	$\dfrac{4}{3}$	$\dfrac{7}{3}$	4	$\dfrac{19}{3}$

(c) $(0,1), (3,4)$

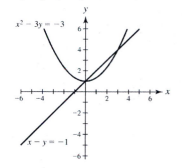

Chapter 14 Sequences and Series

14.1 Sequences

Key Terms

1. infinite sequence

2. nth term; general term

3. finite sequence

Basic Concepts

1. $\dfrac{1}{3}, \dfrac{1}{4}, \dfrac{1}{5}, \dfrac{1}{6}$

2. $-4, 4, -4, 4$

3. $4, 9, 14, 19$

Representations of Sequences

4. $6, 4, 1, 3, 5$

5. $a_n = 65n$ for $n = 1, 2, 3, 4, 5$

n	a_n
1	65
2	130
3	195
4	260
5	325

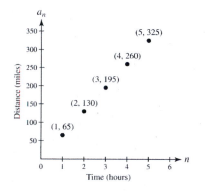

Models and Applications

6. (a) $a_n = 30{,}000(1.1)^{n-1}$

(b)

n	a_n
1	30,000
2	33,000
3	36,300
4	39,930
5	43,923

14.2 Arithmetic and Geometric Sequences

Key Terms

1. arithmetic sequence; common difference

2. geometric sequence; common ratio

3. arithmetic sequence; common difference

Representations of Arithmetic Sequences

1. Yes; $d = -4$

2. No

3. No

4. $a_n = 5n - 6$

5. $a_n = 4 - 3n$

6. $a_{48} = 98$

Representations of Geometric Sequences

7. Yes; $r = 0.05$

8. Yes; $r = -1$

9. No

10. $a_n = -11\left(\dfrac{1}{2}\right)^{n-1}$

11. $a_n = 64\left(\dfrac{1}{4}\right)^{n-1}$

12. $a_{12} = -6144$

Applications and Models

13. (a) $58,000, $61,000, $64,000, $67,000, $70,000; arithmetic

   (b) $a_n = 58,000 + 3000(n-1)$
   $= 3000n + 55,000$

   (c) $a_{10} = 85,000$
   During the tenth year of employment, the employee's salary will be $85,000.

14. (a) 4, 3, 2.25; geometric

   (b) $4(0.75)^{n-1}$

   (c) At the beginning of the sixth day

14.3 Series

Key Terms

1. finite series

2. annuity

3. arithmetic sequence

4. summation notation
   index of summation
   lower limit
   upper limit

5. arithmetic series

6. geometric sequence

Basic Concepts

1. (a) $27,133 + 26,980 + 24,219$
   $+ 24,069 + 25,196 + 20,110$
   $+ 22,337 = 170,044$

   (b) The series represents the total number of internet users from 2001 to 2010.

Arithmetic Series

2. $672,000

3. 360

Geometric Series

4. 547

5. 7.96875

6. $272,615

Summation Notation

7. 385

8. 28

9. 140

10. $\displaystyle\sum_{k=3}^{10} k^3$

11. (a) $\displaystyle\sum_{k=1}^{n} 0.85(0.15)^{k-1}$

   (b) 2 filters

14.4 The Binomial Theorem

Pascal's Triangle

1. $x^5 - 15x^4 + 90x^3 - 270x^2 + 405x - 243$

2. $8r^3 + 12r^2 s + 6rs^2 + s^3$

Factorial Notation and
Binomial Coefficients

3. 120

4. 35

5. $_4C_0 = 1$
    $_4C_1 = 4$
    $_4C_2 = 6$
    $_4C_3 = 4$
    $_4C_4 = 1$

Using the Binomial Theorem

6. $x^6 - 6x^5y + 15x^4y^2 - 20x^3y^3$
    $+15x^2y^4 - 6xy^5 + y^6$

7. $16x^4 - 160x^3 + 600x^2 - 1000x + 625$

8. $15x^4y^2$